# Live Imaging In
# ZEBRAFISH
## Insights into Development and Disease

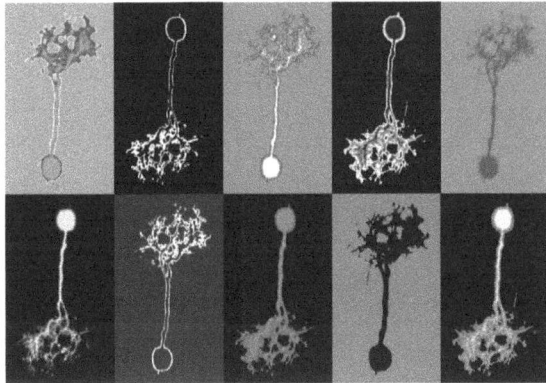

# Live Imaging In
# ZEBRAFISH
## Insights into Development and Disease

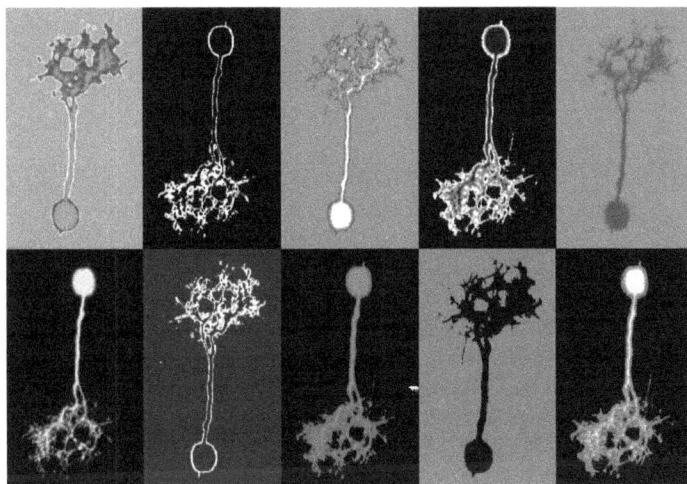

editors

## Karuna **Sampath**
*Temasek Life Sciences Laboratory, Singapore*

## Sudipto **Roy**
*Institute of Molecular and Cellular Biology, Singapore*

**W** **World Scientific**

NEW JERSEY · LONDON · SINGAPORE · BEIJING · SHANGHAI · HONG KONG · TAIPEI · CHENNAI

*Published by*

World Scientific Publishing Co. Pte. Ltd.

5 Toh Tuck Link, Singapore 596224

*USA office:* 27 Warren Street, Suite 401-402, Hackensack, NJ 07601

*UK office:* 57 Shelton Street, Covent Garden, London WC2H 9HE

**British Library Cataloguing-in-Publication Data**
A catalogue record for this book is available from the British Library.

LIVE IMAGING IN ZEBRAFISH
**Insights into Development and Disease**

ISBN-13 978-981-4307-34-5
ISBN-10 981-4307-34-3

Typeset by Stallion Press
Email: enquiries@stallionpress.com

Printed in Singapore.

# Contents

*Preface*                                                                    ix

Chapter 1    Analysis of Branchiomotor Neuron Migration                       1
             in the Zebrafish
             *Petra Stockinger and Carl-Philipp Heisenberg*

             Introduction                                                     2
             Imaging of FBMN Migration *In Vivo*                              5
             Outlook                                                         11
             Acknowledgments                                                 13
             References                                                      13

Chapter 2    Imaging the Nervous System: Insights Into                       17
             Central and Peripheral Glia Development
             *Sarah Kucenas and Bruce Appel*

             Introduction                                                    18
             Central and Peripheral Nervous System                           18
                Development
             *In Vivo* Imaging                                               20
             Central Nervous System                                          21
             Peripheral Nervous System                                       25
             Conclusions                                                     30
             References                                                      31

Chapter 3    Imaging the Cell Biology of Neuronal    35
Migration in Zebrafish
*Martin Distel, Jennifer Hocking and*
*Reinhard W. Köster*

Introduction    36
Neuronal Migration    37
Advances in Technology    43
Cell Biology of Neuronal Migration:    55
   More Questions Than Answers
Conclusions    62
Acknowledgments    63
References    63

Chapter 4    Applications of Fluorescence Correlation    69
Spectroscopy in Living Zebrafish Embryos
*Xianke Shi, Yong Hwee Foo, Vladimir Korzh,*
*Sohail Ahmed and Thorsten Wohland*

Introduction    70
Introduction to FCS    72
Application of FCS in Zebrafish Embryos    83
Conclusion    95
References    96

Chapter 5    Real-Time Imaging of Lipid Metabolism    105
in Larval Zebrafish
*Juliana D. Carten and Steven A. Farber*

Introduction    106
The Clinical Importance of Lipid Therapeutics    106
Limitations of *In Vitro* Lipid Metabolism Studies    107
Zebrafish as Models of Human Physiology    108
   and Disease
Lipid Metabolism is Conserved in Zebrafish    108
Dietary Lipid Metabolism in Zebrafish    109
Real-Time Imaging of Intestinal Lipid    112
   Metabolism: Fluorescent Reporters

Fluorescent Microspheres                          114
Triple Screening: PED6, EnzChek, and
    Microspheres                                  115
Concluding Thoughts                               119
Acknowledgments                                   120
Financial Disclosure                              121
References                                        121

Chapter 6    Live Imaging Innate Immune Cell Behavior     129
             During Normal Development, Wound Healing
             and Infection
             *Chris Hall, Maria Vega Flores, Makoto Kamei,*
             *Kathryn Crosier and Phil Crosier*

             Introduction                                 130
             Mounting Strategies for Live Imaging         132
             Investigating the Physiological Behaviors of 134
                 Myeloid Leukocytes During Normal
                 Development
             Live Imaging the Response of Myeloid         137
                 Leukocytes to Wounding
             Live Imaging the Leukocytic Response to      142
                 Bacterial Infection
             Acknowledgments                              144
             References                                   144

*Index*                                                   149

# Preface

In the past two decades, the zebrafish, *Danio rerio*, has become an established and a widely accepted model for the study of embryonic development, and for understanding the cellular basis of various human diseases. One of the features of this model that has been widely publicized is the optical clarity of the embryos, and the potential for high-resolution microscopy of fixed as well as live samples. However, this salient feature has until recently, been rather underutilized. In order to highlight recent advances using real-time imaging in zebrafish, in this book, we bring together some outstanding examples of state-of-the-art imaging in the context of development, as well as infection and disease.

The first few chapters describe imaging of cell migration in the nervous system, both central as well as peripheral. The use of cell-type specific transgenes, new inducible expression systems, and novel enhancer/promoter cassettes are described in these chapters, with particular emphasis on expression in neuronal cells, oligodendrocytes, and glia. The chapter on fluorescence correlation spectroscopy describes this still very novel methodology, and one for which the zebrafish is particularly suitable. This technique allows real-time imaging and quantification at the single-molecule level, and has the potential to give biophysical insights into the formation of morphogen gradients and ligand-receptor interactions in living embryos. Another area that the zebrafish is now beginning to be appreciated for is its use in understanding organ physiology and function, and this is described in the chapter on imaging digestive physiology. The last chapter covers imaging in the context of infection and wound healing.

We hope that readers working with the zebrafish in areas of developmental biology, cell biology and disease modeling will benefit from the methodologies and tools described in the book, and that it will be a valuable resource for students and researchers alike.

Finally, we thank all the contributors for their time and effort. We also owe special thanks to Ms. Joy Quek of World Scientific for her exceptional patience and hard work.

**Karuna Sampath**
**Sudipto Roy**
*December 2009*

Chapter 1

# Analysis of Branchiomotor Neuron Migration in the Zebrafish

Petra Stockinger and Carl-Philipp Heisenberg

*Max-Planck-Institute of Molecular Cell Biology and Genetics
Dresden, Germany*

## Abstract

Cell migration plays an important role in a wide variety of biological processes. In the developing central nervous system (CNS), many neuronal precursor cells migrate from their site of origin to the final position in which they differentiate. In recent years, studies have often focused on stationary or *in vitro* systems to analyze neuronal migration. However, to understand the cellular and molecular mechanisms of neuronal migration, as well as the malfunction of this process in the diseased brain, it is of great importance to study it in real time in the live intact organism. Advances in microscopy techniques and the development of new dyes and genetically encoded markers have enormously improved *in vivo* time-lapse studies in the last few years. Here, we will describe the zebrafish as a model system to study the migration of a group of motor neurons in the hindbrain. These neurons move from the place where they are born and specified to more caudal regions in the brain. In order to study the cellular dynamics and molecular mechanisms of this process, we are using a stable transgenic line that expresses a green fluorescent protein (GFP) specifically in a subset of the hindbrain motor neurons. Using multiphoton excitation microscopy we can analyze their migration deep within the live tissue over extended periods of time.

*Keywords*:  Zebrafish; Time-Lapse; Neuronal Migration; Central Nervous System; Two-Photon Microscopy.

Correspondence to: Dr. Carl-Philipp Heisenberg, Institute of Science and Technology Austria, Am Campus 1, A-3400 Klosterneuburg, Austria. E-mail: heisenberg@mpicbg.de

1

# Introduction

## Different Types of Neuronal Migration

Throughout development, the nervous system is undergoing major morphogenetic changes and the migration of cells plays an important role in these processes. The occurrence of neuronal cell migration was first noted in the developing cortex by Santiago Ramón y Cajal in the 1890s and was first experimentally addressed in the 1960s and 70s by Pasko Rakic, Mary E. Hatten and others. In the meantime, several groups of migrating neurons and several main types of cell migration in different regions of the developing and adult nervous systems have been identified. Neurons in the cerebral cortex, for example, undergo radial migration from their progenitor zone deep within the cortex to their final laminar position at the surface using radial glia cells, which span the entire cortex, as a substrate.[1] In addition to this radial type of migration, interneurons within different layers of the cerebral cortex have been shown to undergo tangential migration orthogonally to the direction of radial migrating neurons.[2] Finally, chain-like migration of neurons has been observed, e.g. for olfactory neuronal precursors, which are born in the subventricular zone (SVZ) and migrate to the olfactory bulb in close association with each other.[3] Interestingly, neurons adopt a different morphology depending on the type of migration they undergo. Radially migrating neurons typically exhibit a bipolar shape, with opposing leading and trailing process, and only transiently transform into multipolar shapes.[4,5] In contrast, tangentially migrating neurons dynamically change their morphology during their migration.[6] Despite these differences in neuronal morphology and migration, it has been suggested that the molecular and cellular mechanisms underlying neuronal migration are widely conserved.[7]

As of yet, the large majority of studies on neuronal migration have focused on *in vitro* dissociated cell culture systems or *ex vivo* observations in organotypic brain slices. However, to evaluate the relevance of these findings with respect to the *in vivo* situation, it will be critical to study neuronal migration by time-lapse video analysis in the intact organism.

## Caudal Migration of Facial Branchiomotor Neuron (FBMN) in the Hindbrain

Cranial motor neurons control muscles involved in eye, head and neck movements, feeding, as well as speech and facial expression in humans. Mouse, chicken, and more recently zebrafish have been used as experimental model organisms to identify multiple genes that are involved in cranial motor neuron specification, differentiation, migration and axonal guidance.[8,9]

A subset of cranial motor neurons, the branchiomotor neurons (BMN), is specified in certain rhombomeres of the hindbrain. While most BMN will reside in the part of the hindbrain where they are born, neuronal precursors of the facial nerve (nVII), the facial branchiomotor neurons (FBMN), undertake a striking tangential migration along the rostrocaudal axis from rhombomere 4 (r4) to rhombomeres 6 and 7 (r6/7) (Fig. 1A). Despite their migration to r6/7, their axons will still exit the hindbrain at r4 to form components of the facial nerve nVII, which projects to branchial arch 4 and eventually innervates muscles of the head. Several genes have been identified that regulate the specification and migration of these neurons. These include the homeobox genes Hoxa1 and Hoxb1, which are key controllers of hindbrain segmentation and rhombomere identity. Mouse *Hoxa1*, and its zebrafish orthologue *hoxb1b*, are required for the segmental organization of r4, whereas murine *Hoxb1* and zebrafish *hoxb1a* regulate FBMN migration.[10,11] Furthermore, members of the *hedgehog* (*hh*) gene family have been shown to control the expression of the LIM-homeobox gene *Islet1* (*isl1*) and the homeodomain transcription factor Phox2b in postmitotic FBMN in vertebrates.[12] Finally, Tbx20, a member of the T-Box transcription factor family has been suggested to act downstream of *isl1* and *phoxb2*, to regulate a genetic program involved in FBMN migration in mice.[13]

In zebrafish, forward genetic screens have identified a number of mutants that affect caudal migration of FBMN. Interestingly, several of those genes have previously been associated with the non-canonical Wnt/Planar Cell Polarity (PCP) pathway,[12,14–17] which plays important

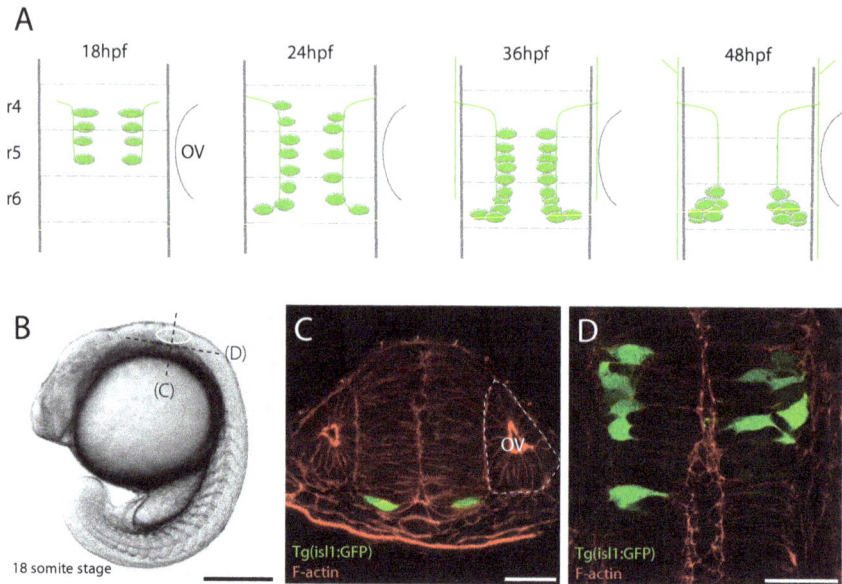

**Fig. 1** Facial branchiomotor neurons (FBMN) in the zebrafish hindbrain undergo caudal migration during development. (**A**) Schematic drawing of FBMN migration in zebrafish. Dorsal views of the zebrafish hindbrain at the corresponding hours post-fertilization (hpf). The FBMN and their axons are shown in green and broken lines indicate rhombomeric boundaries (modified from Ref. 15). (**B**) DIC picture of a live 18 somite-stage embryo (18 hpf; lateral view). The otic vesicle (OV) is outlined in white and the plane of section for (C, D) is indicated with a dashed line. (**C**) A transverse section through the hindbrain at rhombomere 5 marked by the OV, shows the position of the FBMN close to the ventral midline (floorplate). (**D**) Coronal section through the hindbrain at the level of the FBMN migratory route. Anterior is at the top and posterior at the bottom. Filamentous actin (C, D) is labeled by phalloidin (red) and FBMN are labeled by the transgenic line *Tg*(*isl1:GFP*) (green). Scale bars: (B) 250 $\mu$m; (C, D) 30 $\mu$m.

roles in the planar organization of epithelial structures and morphogenetic movements in both vertebrates and invertebrates.[18–20] Genetic mosaic analyses have revealed that in the hindbrain, Wnt/PCP genes are predominantly required within the neuroepithelium surrounding the FBMN.[14–16] Thus, it has been speculated that Wnt/PCP signaling provides environmental cues involved in modulating the movement of FBMN from r4 to r6/7. Much less, however, is known about factors that act within the FBMN to regulate their movement. An exception is the Wnt/PCP pathway component Prickle1b, which is expressed within the FBMN and cell-autonomously required for their migration.[21]

Importantly, until now the analysis of FMBN migration has focused on the identification of new genes and their function has predominantly been studied through stationary methods such as *in situ* hybridization or antibody staining. However, in order to understand the dynamics of FBMN migration and to identify the underlying molecular and cellular mechanisms, high-resolution time-lapse analysis of migrating motor neurons is required. In the following we will describe how two-photon (2P) live imaging can help in studying FBMN migration within the zebrafish hindbrain.

## Imaging of FBMN Migration *In Vivo*

### Zebrafish as a Model System to Study FBMN Migration

Why study FBMN migration in zebrafish? The zebrafish is a widely used vertebrate model system, which produces large numbers of embryos that undergo rapid development outside the mother. Because zebrafish embryos are transparent, cell behaviors and interactions can be easily monitored in live embryos using different microscopy techniques (Fig. 1B). Moreover, the availability of stable transgenic lines, such as the *Tg*(*isl1:GFP*) transgene to label FBMN,[22] allows the ready visualization of distinct populations of cells in the living organism. Finally, zebrafish are highly amenable for reverse and forward genetic tools, facilitating the functional analysis of genes involved in different developmental processes.

So far high-resolution live imaging of FBMN migration in the zebrafish has been sparse mainly due to experimental limitations. The FBMN are located relatively deep — around 100 $\mu$m — within the embryo on the ventral side of the neural tube, in close proximity to the floorplate (Fig. 1C). It is therefore difficult to image them with standard laser scanning confocal microscopy. One way to circumvent this problem has been to image hindbrain explants in culture,[23] however, it is not yet clear how much this culture system interferes with the migration of FBMN. Thus, in order to image FBMN at high resolution within the intact organism, we have focused on adapting and developing microscopy techniques, such as 2P microscopy, suitable to monitor cells deep within the tissue over extended periods of time.

## 2P Microscopy as a Tool to Study FBMN Migration

The expression of fluorescent proteins is usually analyzed with standard laser scanning confocal microscopes to acquire high-resolution optical images. A key feature of confocal microscopy is its ability to produce in-focus images of thick specimens, a process known as optical sectioning. However, mainly due to the scattering of light through the tissue, the depth of observation is limited and it is thus difficult to study cells that are located deep in a tissue and/or have relatively low levels of fluorescence. Furthermore, confocal imaging can cause photobleaching and subsequent phototoxicity, limiting its use for time-lapse imaging of biological specimen. Multiphoton Fluorescence Microscopy is a relatively novel imaging technique in cell biology which provides two major advantages to regular confocal imaging.[24] First, it allows the generation of high-resolution three-dimensional images deep within live samples. This is because multiphoton microscopy uses higher frequency light, which can penetrate deeper into the tissue.[25] Secondly, photobleaching and phototoxicity are comparably low as multiphoton excitation only occurs at the focal plane[26] instead of in the entire light path through the sample as is the case in standard confocal microscopy.[27–29] There are also some limitations to using multiphoton microscopy. The image resolution obtained with 2P excitation is not better than that achieved in a well aligned confocal microscope and in thin specimens, photobleaching in the focal plane is higher as compared to conventional laser scanning microscopy.

## Aspects of FBMN Migration Addressable by 2P Microscopy

Previous time-lapse analysis using standard laser scanning confocal microscopy has already provided some insight into the cellular basis of FBMN migration.[16,23] It has been shown that FBMN have posteriorly-biased expansion and anteriorly-biased retraction zones, resulting in their effective posterior translocation. Moreover, analysis of FBMN migration in Wnt/PCP mutant embryos has provided evidence that mutant FBMN move more slowly and in random directions, and it has been speculated that this migration phenotype is caused by defects in the stability and/or polarization of their cell protrusions. However, to verify these observations, the spatial and temporal resolution of imaging FBMN migration and morphology still needs to be improved considerably. Questions regarding FBMN development that can be specifically addressed by 2P microscopy include: How do FBMN transform from a stationary epithelial neuronal precursor cell into a mobile one? How do they migrate through the epithelium? What is the substrate on which FBMN migrate? How do they interact with each other and/or the surrounding epithelium? What are the intrinsic and extrinsic factors determining FBMN migration? How and when is their initial polarity established, instructing them where to form dendrites and axons?

## 2P Microscopy Setup to Study FBMN Migration

In order to image the migration of FBMN in the live animal, a transgenic line expressing GFP under the control of a promoter/enhancer region of the transcription factor islet1 can be used.[22] This line specifically labels BMN after their final cell division and outlines BMN cell bodies and axonal processes including the peripheral branches within the muscles. By using this line, different aspects of FBMN development such as the initiation of migration, the type of movement, and the outgrowth and pathfinding of motor axons can be easily studied. To image the shape and behavior of the surrounding neuroepithelium during neuronal migration, the cell membrane of all cells in the embryo can be labeled by ubiquitously expressing a membrane-bound version of GFP (EGFPCAAX) or using a transgenic line (e.g. *Tg(β-actin:HRAS-EGFP)*), which drives ubiquitous expression of membrane-bound GFP.[30] Embryos are usually dechorionated at 18 hours post-fertilization, when the first FBMN are about to migrate caudally, and mounted within a self-assembled imaging chamber in low-melting point agarose to immobilize them (Fig. 2A). To further inhibit embryo motility, the

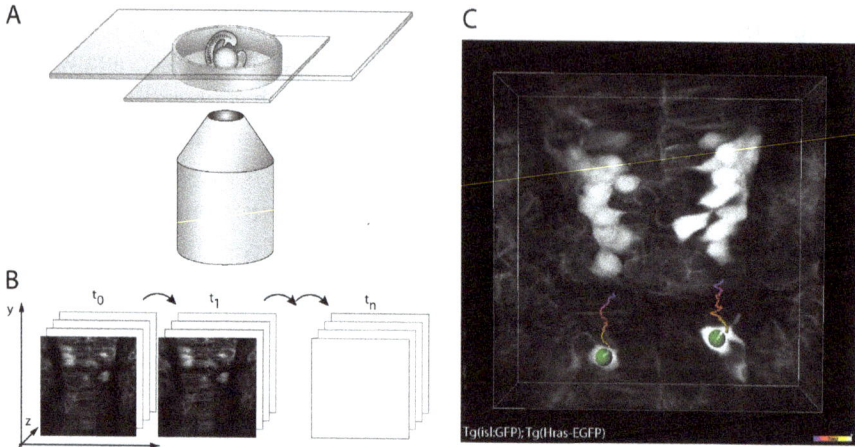

**Fig. 2** Two-photon (2P) microscopy setup and image data analysis. **(A)** Mounting of embryos for imaging. Mounting chamber and embryo orientation used to image embryos on an inverted 2P microscope. The mounting chamber consists of a cover-slip attached to a glass ring in which the embryo is placed in agarose medium. The region of the embryo that is going to be imaged (here the region of the hindbrain) should face the coverslip. **(B)** Image acquisition of a 4D data with *xyz* stacks obtained over *n* timepoints (*t*). **(C)** Image visualization, segmentation and analysis using Imaris software. The obtained 4D data sets are used for semi-automated tracking of cell shape and cell migration. The surfaces of the segmented objects are obtained and their centers are tracked over time. The final positions of the tracked FBMN are indicated by green dots and the movement of their center is visualized via color-coded tracks (color as a function of time).

anesthetic Tricaine is added to the agarose-containing mounting medium. For monitoring the movement and morphology of FBMN over time, we acquire a four-dimensional (4D) data set with high spatial and temporal resolution. Usually an XY resolution of $512 \times 512$ pixels with a scanning speed of 166 lines per seconds (lps) and a stack size of around 25 slices is sufficient (Fig. 2B). Since the speed of zebrafish development, and, as a result of this, of FBMN migration is temperature-dependent, we carefully control the imaging temperature through the use of objective heaters and imaging chambers.

For analysis of the 4D data set, we use several types of imaging software. Volocity (Improvision) and Imaris (Bitplane) are two commonly used commercial programs, which allow fast and easy handling of large 4D data sets. Alternatively, the free software platform ImageJ (http://rsb.info.nih.gov/ij/) provides excellent plugins and recordable macros, which facilitate manual tracking of objects in 4D image sequences and subsequent analysis of various movement and morphology parameters (Fig. 2C).

## Advantages of 2P Microscopy to Study FBMN Migration

By using multiphoton time-lapse imaging of *islet1-GFP* transgenic zebrafish embryos, we have begun to visualize and analyze the movement of FBMN from their origin in r4 through to their incorporation into r6/7. We have found that before FBMN begin to migrate, they are bipolar elongated cells, which are connected to both the apical and basal sides of the neuroepithelium (Fig. 3A). To initiate migration, the first migrating FBMN retract their apical side and round up close to the

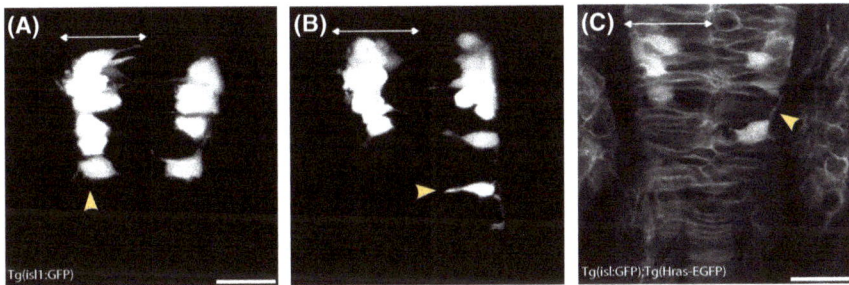

**Fig. 3** FBMN migration. **(A)** Before migration, FBMN are bipolar and connected with the apical and basal side of the neural tube. The double arrow (A–C) indicates the width of the neuroepithelium (NE) with the apical side facing the midline and the basal side to the left. The most posterior cells start to migrate first by rounding up at the basal side of the NE and sending out protrusions towards the posterior (arrowhead). Images show a z-projection where FBMN are labeled by the transgenic line *Tg(isl1:GFP)*. **(B)** During migration, FBMN also extend protrusions towards the midline (arrowhead) and become bipolar in shape, while keeping long processes towards the posterior. **(C)** Migrating FBMN also stay connected with r4 via long processes (arrowhead). Image shows a single z-plane where all membranes are labeled with membrane bound EGFP (*Tg(Hras-EGFP)*). Scale bars: (A–C) 30 μm.

outer basal side of the neuroepithelium. They then become migratory by dynamically extending and retracting protrusions posteriorly towards the direction of their migration as well as medially towards the midline (Figs. 3A and 3B). While migrating posteriorly, they are leaving behind a long axonal process, which exits the neural tube at r4 (Fig. 3C).

How can these high-resolution time-lapse movies help in the analysis of FBMN migration? First of all, they provide the material for a detailed quantitative description of the migration process *in vivo*. Using appropriate image-analysis software (see above), we can either manually or semi-automatically track individual migrating FBMN over time in 3D and determine different parameters of migration such as movement speed, directionality and coherence. Furthermore, we can determine different morphological features of single migrating FBMN, including cell elongation, polarization and protrusion formation. Eventually, morphological features can be correlated with migratory parameters, providing information about the potential contribution of changes in cell morphology to different aspects of cell migration.

2P time-lapse movies will also be very useful to elucidate how intracellular structures, such as components of the cytoskeleton and trafficking machinery, are reorganized in migrating FBMN. Such components can be visualized in FBMN by expressing fluorescently tagged fusion proteins. Eventually, these studies will provide information about dynamic changes of intracellular components during migration and thus their potential function in this process.

Finally, 2P time-lapse movies will be an indispensable tool for the analysis of gene function in FBMN development. In general, gene function can be modulated by both gain-of-function and loss-of-function approaches. Gain-of-function is usually achieved by the over-expression of a gene product via injection of DNA or synthetic transcribed *m*RNA, while loss-of-function relies on the use of antisense morpholino oligonucleotides to silence translation of specific proteins or forward genetic mutagenesis approaches to identify loss-of-function mutations. 2P live analysis of the FBMN migration process in wild type embryos and embryos with reduced/increased gene function can then be used to reveal the requirement of specific genes during the migration process. Furthermore, to distinguish between cell autonomous and cell-non-autonomous gene function in FBMN migration, genetic mosaic embryos can be analyzed. Usually, these are generated by transplanting cells at an early stage from a donor embryo into a host embryo of a different genotype. 2P live imaging of such mosaic embryos can then be used to determine whether a gene is required within FBMN themselves and/or in the surrounding tissues.

# Outlook

## Alternative Microscopy Techniques

Two advantages of multiphoton microscopy are its inherent ability to provide high resolution three-dimensional images at significant depths and the possibility to do this in live tissue with minimal sample damage. However, there are some drawbacks to this imaging method. First of all, it is still too slow to study dynamic cell behaviors such as the extension and retraction of cell protrusions and the movements of subcellular structures. In order to analyze these processes, faster microscopic techniques, such as spinning disc confocal microscopy, will be very useful. This technique is based on standard confocal laser scanning microscopy, but in contrast to single beam scanning, the parallelized approach of multi-beam scanning allows much faster imaging.[31] Although perfectly suited for fast *in vivo* imaging, spinning disk confocal microscopy still has the same disadvantages as standard confocal imaging with respect to limited sample depth and bleaching of the fluorophore.

Other recent microscopy developments to investigate large specimens with a high and isotropic three-dimensional resolution are light sheet-based technologies. There, optical sectioning is achieved by illuminating the sample along a separate optical path orthogonal to the detection axis. For example in selective plane illumination microscopy (SPIM), the excitation light is focused by a cylindrical lens to a sheet of light that illuminates only the focal plane of the detection optics, so that no out-of-focus fluorescence is generated.[32-34] Recently, a new implementation of light sheet based microscopy has been developed.

Digital scanned laser light sheet fluorescent microscopy (DSLM) allows quantitative imaging with a higher imaging quality through better illumination efficiency.[35] In general, these setups are well suited for large samples to study features that require high resolution imaging over long periods of time. They might also be useful to study FBMN migration in the zebrafish hindbrain, as they combine the advantage of deep 3D imaging by low excitation energy with relatively fast imaging speeds. However whether SPIM or DSLM provide images with sufficiently high subcellular resolution to study the migration behavior of FBMN remains to be addressed.

## Spatial and Temporal Gene and Cell Manipulations

Apart from these new microscopy techniques, the development of tools to control gene expression, manipulate gene function, and release bioactive compounds in individual cells in a temporarily controlled manner, would represent important improvements for the analysis of FBMN migration. For example, photoactivatable fluorescent proteins that exhibit pronounced light-induced spectral changes by near-infrared femtosecond laser pulses can be used to label individual FBMN or, potentially, the subcellular structures they contain.[36,37] Another possibility would be to use multiphoton excitation to release and thus activate bioactive compounds, such as morpholino antisense oligonucleotides, from a molecular cage in individual FBMN.[38] Multiphoton excitation is particularly suitable for the localized release of caged compounds, since excitation is confined to a small volume within the focal plane.[26] Multiphoton excitation can also be used for cell ablation studies,[39] providing insight into the function of individual FBMN and/or the requirement of surrounding tissues for FBMN migration. Taking advantage of the spatially confined activity of multiphoton excitation, single cells can be ablated with only minimal damage to neighboring cells and tissues. The ability to regulate gene expression in a cell-specific and temporally controlled manner constitutes a powerful tool to analyze gene function in specific biological processes such as FBMN migration. The most commonly used genetic tools to achieve this are the GAL4/UAS system established in *Drosophila*[40] and other methods taking advantage of site-specific recombinases such as the Cre/loxP system, which are commonly used in mice.[41] In principle, these systems are available in zebrafish but have not yet been rigorously tested in the context of FBMN migration.[42,43] Alternatively, multiphoton laser excitation would constitute a very suitable way to activate gene expression by heating single cells containing genes under the control of heat shock-inducible promoters.[44] In the future, the combination of improved and new imaging techniques and spatially and temporarily regulated cell and gene manipulations will allow the study of neuronal migration in greater detail.

## Concluding Remarks

FBMN migration in the zebrafish hindbrain constitutes a promising assay system to study the molecular and cellular mechanisms underlying neuronal cell migration in development. Moreover, the use of 2P microscopy to analyze FBMN migration has turned out to be extremely useful to study their behavior on the subcellular levels deep within the whole live developing embryo.

## Acknowledgments

We are grateful to H. Okamoto and J. Topczewski for providing us with the Tg(*isl1:GFP*) and the Tg(*β-actin:HRAS-EGFP*) fish line respectively. We also thank J. Peychl for advice and assistance with 2PE microscopy and G. Junghanns, E. Lehmann, and J. Hückmann for help with the fish care. We thank K. Margitudis for help with figures and G. Soete for critical reading of the manuscript.

## References

1. Marin O and Rubenstein JLR. Cell migration in the forebrain. *Annu Rev Neurosci* 26:441–483, 2003.
2. Marin O and Rubenstein JLR. A long, remarkable journey: tangential migration in the telencephalon. *Nat Rev Neurosci* 2:780–790, 2001.

3. Lois C, García-Verdugo J-M and Alvarez-Buylla A. Chain migration of neuronal precursors. *Science* 271:978–981, 1996.
4. LoTurco JJ and Bai J. The multipolar stage and disruptions in neuronal migration. *Trends Neurosci* 29:407–413, 2006.
5. Tabata H and Nakajima K. Multipolar migration: the third mode of radial neuronal migration in the developing cerebral cortex. *J Neurosci* 23:9996–10001, 2003.
6. Hatten ME. Central nervous system neuronal migration. *Annu Rev Neurosci* 22:511–539, 1999.
7. Marín O Valdeolmillos M and Moya F. Neurons in motion: same principles for different shapes? *Trends Neurosci* 29:655–661, 2006.
8. Guthrie S. Patterning and axon guidance of cranial motor neurons. *Nat Rev Neurosci* 8:859–871, 2007.
9. Anand C. Turning heads: development of vertebrate branchiomotor neurons. *Dev Dyn* 229:143–161, 2004.
10. Trainor PA and Krumlauf R. Patterning the cranial neural crest: hinbrain segmentation and hox gene plasticity. *Nat Rev Neurosci* 1:116–124, 2000.
11. McClintock JM, Kheirbek MA and Prince VE. Knockdown of duplicated zebrafish hoxb1 genes reveals distinct roles in hindbrain patterning and a novel mechanism of duplicate gene retention. *Development* 129: 2339–2354, 2002.
12. Bingham S, Higashijima S, Okamoto H and Chandrasekhar A. The zebrafish trilobite gene is essential for tangential migration of branchiomotor neurons. *Dev Biol* 242:149–160, 2002.
13. Song MR, Shirasaki R, Cai CL, Ruiz EC, Evans SM, Lee SK and Pfaff SL. T-Box transcription factor Tbx20 regulates a genetic program for cranial motor neuron cell body migration. *Development* 133:4945–4955, 2006.
14. Wada H, Tanaka H, Nakayama S, Iwasaki M and Okamoto H. Frizzled3a and Celsr2 function in the neuroepithelium to regulate migration of facial motor neurons in the developing zebrafish hindbrain. *Development* 133:4749–4759, 2006.
15. Wada H, Iwasaki M, Sato T, Masai I, Nishiwaki Y, Tanaka H, Sato A, Nojima Y and Okamoto H. Dual roles of zygotic and maternal Scribble1 in neural migration and convergent extension movements in zebrafish embryos. *Development* 132:2273–2285, 2005.
16. Jessen JR, Topczewski J, Bingham S, Sepich DS, Marlow F, Chandrasekhar A and Solnica-Krezel L. Zebrafish trilobite identifies new roles for Strabismus in gastrulation and neuronal movements. *Nat Cell Biol* 4:610–615, 2002.

17. Carreira-Barbosa F, Concha ML, Takeuchi M, Ueno N, Wilson SW and Tada M. Prickle 1 regulates cell movements during gastrulation and neuronal migration in zebrafish. *Development* 130:4037–4046, 2003.
18. Zallen JA. Planar polarity and tissue morphogenesis. *Cell* 129:1051–1063, 2007.
19. Seifert JRK and Mlodzik M. Frizzled/PCP signalling: a conserved mechanism regulating cell polarity and directed motility. *Nat Rev Genet* 8:126–138, 2007.
20. Simons M and Mlodzik M. Planar cell polarity signaling: from fly development to human disease. *Annu Rev Genet* 42:517–540, 2008.
21. Rohrschneider MR, Elsen GE and Prince VE. Zebrafish Hoxb1a regulates multiple downstream genes including prickle1b. *Dev Biol* 309:358–372, 2007.
22. Higashijima S, Hotta Y and Okamoto H. Visualization of cranial motor neurons in live transgenic zebrafish expressing green fluorescent protein under the control of the islet-1 promoter/enhancer. *J Neurosci* 20:206–218, 2000.
23. Bingham SM, Toussaint G and Chandrasekhar A. Neuronal development and migration in zebrafish hindbrain explants. *J Neurosci Methods* 149:42–49, 2005.
24. Denk W, Strickler JH and Webb WW. Two-photon laser scanning fluorescence microscopy. *Science* 248:73–76, 1990.
25. Oheim M, Beaurepaire E, Chaigneau E, Mertz J and Charpak S. Two photon microscopy in brain tissue: parameters influencing the imaging depth. *J Neurosci Methods* 111:29–37, 2001.
26. Zipfel WR, Williams RM and Webb WW. Nonlinear magic: multiphoton microscopy in the biosciences. *Nat Biotechnol* 21:1369–1377, 2003.
27. Helmchen F and Denk W. Deep tissue two-photon microscopy. *Nat Methods* 2:932–940, 2005.
28. Diaspro A, Chirico G and Collini M. Two-photon fluorescence excitation and related techniques in biological microscopy. *Q Rev Biophys* 38:97–166, 2005.
29. Svoboda K and Yasuda R. Principles of two-photon excitation microscopy and its applications to neuroscience. *Neuron* 50:823–839, 2006.
30. Cooper MS, Szeto DP, Sommers-Herivel G, Topczewski J, Solnica-Krezel L, Kang HC, Johnson I and Kimelman D. Visualizing morphogenesis in transgenic zebrafish embryos using BODIPY TR methyl ester dye as a vital counterstain for GFP. *Dev Dyn* 232:359–368, 2005.

31. Nakano A. Spinning-disk confocal microscopy — a cutting-edge tool for imaging of membrane traffic. *Cell Struct Funct* 27:349–355, 2002.
32. Huisken J, Swoger J, Del Bene F, Wittbrodt J and Stelzer EHK. Optical sectioning deep inside live embryos by selective plane illumination microscopy. *Science* 305:1007–1009, 2004.
33. Verveer PJ, Swoger J, Pampaloni F, Greger K, Marcello M and Stelzer EH. High-resolution three-dimensional imaging of large specimens with light sheet based microscopy. *Nat Methods* 4:311–313, 2007.
34. Huisken J and Stainier DY. Even fluorescence excitation by multidirectional selective plane illumination microscopy (mSPIM). *Opt Lett* 32:2608–2610, 2007.
35. Keller PJ, Schmidt AD, Wittbrodt J and Stelzer EH. Reconstruction of zebrafish early embryonic development by scanned light sheet microscopy. *Science* 322:1065–1069, 2008.
36. Kozlowski DJ and Weinberg ES. Photoactivatable (caged) fluorescein as a cell tracer for fate mapping in the zebrafish embryo. *Methods Mol Biol* 135:349–355, 2000.
37. Hatta K, Tsujii H and Omura T. Cell tracking using a photoconvertible fluorescent protein. *Nat Protoc* 1:960–967, 2006.
38. Shestopalov IA, Sinha S and Chen JK. Light-controlled gene silencing in zebrafish embryos. *Nat Chem Biol* 3:650–651, 2007.
39. Supatto W, Debarre D, Moulia B, Brouzes E, Martin JL, Farge E and Beaurepaire E. *In vivo* modulation of morphogenetic movements in Drosophila embryos with femtosecond laser pulses. *Proc Natl Acad Sci USA* 102:1047–1052, 2005.
40. Phelps CB and Brand AH. Ectopic gene expression in Drosophila using GAL4 system. *Methods* 14:367–379, 1998.
41. Orban PC, Chui D and Marth JD. Tissue- and site-specific DNA recombination in transgenic mice. *Proc Natl Acad Sci USA* 89:6861–6865, 1992.
42. Thummel R, Burket CT, Brewer JL, Sarras Jr, MP, Li L, Perry M, McDermott JP, Sauer B, Hyde DR and Godwin AR. Cre-mediated site-specific recombination in zebrafish embryos. *Dev Dyn* 233:1366–1377, 2005.
43. Halpern ME, Rhee J, Goll MG, Akitake CM, Parsons M and Leach SD. Gal4/UAS transgenic tools and their application to zebrafish. *Zebrafish* 5:97–110, 2008.
44. Shoji W and Sato-Maeda M. Application of heat shock promoter in transgenic zebrafish. *Dev Growth Differ* 50:401–406, 2008.

Chapter 2

# Imaging the Nervous System: Insights Into Central and Peripheral Glia Development

Sarah Kucenas

*Department of Biology, University of Virginia*
*Charlottesville, VA, USA*

Bruce Appel

*Department of Pediatrics, University of Colorado-Denver*
*Aurora, CO, USA*

## Abstract

Formation of a functional nervous system requires the coordinated interaction of neurons and glial cells. These cells often migrate considerable distances from their origins, recognize each other and form intimate and very specific physical connections. The highly dynamic behaviors of neural cells during development are not adequately captured by standard experimental approaches of fixed tissue analysis and *in vitro* methods. To facilitate direct observation of neural cells within intact embryos, we created transgenic zebrafish that express fluorescent proteins under control of cell type-specific promoters and developed methods for long term, *in vivo*, time-lapse imaging. In this chapter, we present a brief overview of central and peripheral nervous system development and describe a few of our

Correspondence to: Dr. Sarah Kucenas, Ph.D., Department of Biology, University of Virginia, Gilmer Hall, Room 275, P.O. Box 400328, Charlottesville, VA 22904-4328, USA, E-mail: sk4ub@virginia.edu studies that have revealed new cell origins and behaviors and our investigations of gene functions that direct glial cell specification and differentiation.

*Keywords:* Glia; Zebrafish; Neural Development; Live Imaging; CNS; PNS.

## Introduction

The growth of an efficient and functional nervous system requires that the developmental programs of neurons and glia are coordinated. Cells of both classes must be specified at the correct place and time and in appropriate numbers. Subsequently, many of the newly specified cells must initiate long-distance migrations and undergo morphological and functional differentiation, which ultimately is necessary for formation of neural circuits. Work aimed at understanding nervous system development has clearly demonstrated that neural cell specification, migration and differentiation depend on dynamic interactions between cells.[1] Although current techniques utilizing cell culture or fixed tissue preparations have provided key insights, they fall short of describing how cells interact and how these interactions result in a functional unit. A full appreciation of cell interactions and the genetic control of these behaviors can only be achieved by investigating them in an intact, live organism.

As a model organism, zebrafish provides a unique combination of *in vivo*, time-lapse imaging and genetics. The creation of many neural-specific transgenic lines, the acquisition of mutant lines with phenotypes that affect nervous system components and advanced *in vivo* imaging techniques has created the opportunity to observe the behaviors of neural cells and their interactions with each other during nervous system development.

## Central and Peripheral Nervous System Development

For animals to develop, survive and reproduce, an exquisite array of neural cells must be specified and maintained. The simplicity and accessibility of the central and peripheral nervous systems (CNS and PNS, respectively) of zebrafish makes them attractive models to address questions of vertebrate nervous system development. Our lab is interested in using the spinal cord and adjoining PNS components, including motor nerves, as models to investigate glial cell development and neuronal-glial and glial-glial interactions.

The earliest steps of spinal cord development in zebrafish embryos occur via a morphogenetic mechanism called secondary neurulation, whereby the neural plate condenses to form a solid rod. Neurulation begins after gastrulation, at approximately the two-somite stage [ten hours-post-fertilization (hpf)], with the thickening of the medial neural plate by cell movements directed toward the midline.[2,3] In zebrafish, like other anamniotes, neurogenesis occurs in two distinct phases with the first phase, primary neurogenesis, initiated during early neurulation. During this phase, three classes of primary neurons, primary motor neurons, primary interneurons and mechanosensory Rohon-Beard neurons, are specified. These neurons initially occupy distinct, bilateral, medial, intermediate and lateral longitudinal columns within the neural plate. However, as the neural plate undergoes convergence and extension movements that first create the neural keel and subsequently the solid neural cord, these bilateral, longitudinal columns are translated into dorsal-ventral domains within the cord.[2,3] Sensory neurons, which are specified in the lateral edges of the neural plate, occupy the most dorsal positions of the neural cord whereas motor neurons, which are specified within the medial neural plate, ultimately become positioned in the ventral cord. Interneurons, specified in the intermediate regions of the neural plate, reside between motor and sensory neurons. The final steps of neurulation ultimately result in a mature neural tube with a lumen created via cavitation of the solid neural cord.[2,3] Secondary neurogenesis begins just after neural tube cavitation at approximately 17 hpf, continues until 1.5 days post-fertilization (dpf) and adds many motor neurons and interneurons to the neural tube.[4]

Some of the first neurons to begin axonogenesis during nervous system development are the primary motor neurons of the spinal cord. Between 16 and 18 hpf, they begin to extend axons out of the spinal cord and into the PNS at iterated positions along the anterior-posterior axis of the embryo called motor exit points (MEPs).[5] These axons are some of the first components of the newly forming PNS and are immediately associated with Schwann cells, the myelinating glia cell of the PNS, as they exit. Schwann cells, a neural crest derivative, arise from the dorsal surface of the neural tube and migrate ventrally via the medial pathway between the neural tube and somite during neurulation and ultimately become associated with pioneering primary motor axons.[6,7] Once in the PNS, motor axon growth cones pioneer into the muscle and respond to cues in the environment that guide them towards their target muscle field and they ultimately make specialized synapses known as neuromuscular junctions with individual muscle fibers.[4,8]

During the first day of development, neurulation and neurogenesis create a functional, yet still growing nervous system that allows embryos to react to their environment. Over the next 48 hours, the complexity of the CNS and PNS increases with the specification and differentiation of additional classes of neurons. Also, glial cells of the CNS, including astrocytes, oligodendrocytes and radial glia, are specified, differentiate and interact with other neural cells in the spinal cord to ensure the creation of functional, efficient neural circuits.

With the incredible diversity of cells specified during development, a mature, functional nervous system requires that all developmental programs of neurons and glia be precisely orchestrated to ensure that appropriate cell-cell interactions are made. It is the nature of these dynamic interactions and the mechanisms that govern them that have remained elusive. Specifically, we are interested in investigating questions including: How do glial cells interact with each other and axons in both the CNS and PNS? How is an even distribution of myelin along axonal segments in the CNS and PNS achieved? And how do glial cells in the CNS and PNS respond to damage or cell loss? The most powerful and straightforward way to answer these questions is by directly observing the cells in an intact model system using time-lapse imaging.

## *In Vivo* Imaging

One of the unique advantages to zebrafish embryos and larvae for studying development of the nervous system is their transparency. This feature, in combination with small embryo size, fast development and the ease with which embryos can be kept alive, allows for high-resolution, time-lapse imaging of cells in the CNS and PNS in intact animals. However, obtaining high-quality images of cell behaviors within the complex, three-dimensional structure of the nervous system can be challenging. To capture the best images possible, we used confocal microscopy. The most common type of confocal microscope scans a single, focused, high intensity laser beam across a region and collects emitted light through a small pinhole aperture.[9] Although these microscopes can produce exceptional images, exposure to the high intensity laser can be toxic to cells and the organism as a whole. In contrast, spinning disk confocal microscopes illuminate the embryo evenly by delivering the laser light through multiple pinholes on a rotating disk.[10] Because multiple beams of the laser pass rapidly and sequentially over the same point of the embryo during exposure, much less intense illumination is needed to produce emitted light,[10] and therefore, allows for extended periods of imaging without damage to the embryo. For our purposes, we use spinning disk confocal microscopy to routinely collect high resolution images of the CNS and PNS every two to three minutes for up to 48 hours without evidence of phototoxicity to the embryo. One important limitation, however, is the fixed pinhole size. In this situation, the numerical aperture and magnification of the objective determines the depth to which we can image and out-of-focus fluorescence has the potential to decrease image quality. However, in our lab, we achieve excellent results with minimal contamination of unfocused fluorescence when using a spinning disk confocal microscope for imaging the spinal cord and peripheral motor nerves of embryos and larvae.

## Central Nervous System

### Behaviors and Interactions of Oligodendrocyte Progenitor Cells

Over the last two decades, the zebrafish has become a valuable model to help dissect the molecular mechanisms that mediate the specification of neurons and glia. However, what have been lacking are techniques that allow us to investigate how these cells interact after specification and what happens when these interactions are perturbed.

The recent production of transgenic lines that drive expression of fluorescent reporter proteins in neurons and glia, including *Tg(olig2:egfp)*,[11] *Tg(nkx2.2a:megfp)*,[12,13] *Tg(ngn1:egfp)*,[14] *Tg(sox10 (7.2):mrfp)*,[13] *Tg(alx:gfp)*,[15] *Tg(HuC:kaede)*,[16] *Tg(gfap:gfp)*[17] and *Tg(isl1:gfp)*,[18] coupled with new *in vivo*, time-lapse imaging techniques, have given us the opportunity to visualize these intimate and dynamic interactions.

Our lab is currently interested in what mechanisms determine the number and distribution of oligodendrocytes in the spinal cord. In the vertebrate CNS, the rapid propagation of action potentials along large-diameter axons requires that these axons be wrapped in myelin, the fatty sheath of lipids and lipoproteins that is produced by oligodendrocytes[19–21]. During development, distinct populations of ventral spinal cord neural precursors give rise to oligodendrocyte progenitor cells (OPCs).[22,23] After specification, ventrally born OPCs divide and migrate to become uniformly distributed throughout the spinal cord.[20] However, little is known about the mechanisms that regulate this distribution. Previous work in mouse implicated the role of limiting amounts of neuronal and astrocyte-derived growth factors, including platelet-derived growth factor (PDGF), in OPC division and migration.[24,25] To further investigate how oligodendrocytes become uniformly distributed in the spinal cord, we utilized time-lapse imaging of transgenic embryos and focused on the behaviors of OPCs with each other and the surrounding environment.

Specifically, we utilized the *Tg(nkx2.2a:megfp)* line, which labels a subset of ventrally originating OPCs with a membrane-tethered enhanced green fluorescent protein[12] because we can visualize not only the soma of the cells, but their cellular processes as well. Between 36 hpf, when OPCs are specified and migrate out of the ventral spinal cord, and 72 hpf, migratory OPCs have highly branched membrane processes that often appear to point in the direction of the cells' movements (Figs. 1a and 1b).[12] However, we know very little about the mechanism that mediates how an OPC chooses its trajectory. One possibility is their membrane processes respond to cues in the environment. Currently, we do not know the nature of these molecular cues, but further

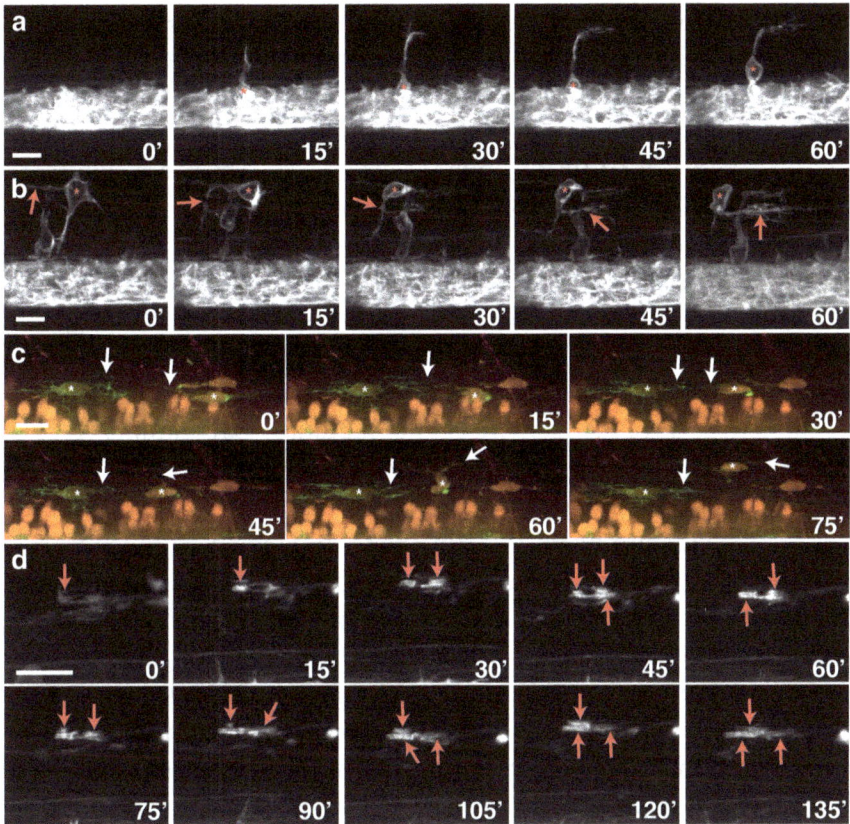

**Fig. 1** *In vivo*, time-lapse imaging reveals dynamic behaviors of OPCs in the spinal cord. (a and b) Frames captured from a 24-hour time-lapse sequence of a *Tg(nkx2.2a:megfp)* embryo beginning at 50 hpf. (a) At 50 hpf (0-min time point), a long EGFP⁺ process emerged from the ventral spinal cord, followed by a cell body (red asterisk). (b) At approximately 64 hpf (0-min time point), an EGFP⁺ OPC (red asterisk) is observed following its leading membrane process (red arrow). (c) Frames captured from a 24-hour time-lapse sequence of a *Tg(nkx2.2a:megfp); Tg(olig2:dsred2)* embryo beginning at 52 hpf. At approximately 60 hpf (0-min time point), membrane processes (white arrows) of two OPCs (white asterisks) interact and processes are then observed retracting and one OPC changes direction and directs process extensions in a different direction. (d) Frames from a 24-hour time-lapse sequence of a *Tg(nkx2.2a:megfp)* embryo beginning at 50 hpf. At approximately 70 hpf (0-min time point), OPC membrane processes are observed ensheathing CNS axons (red arrows). Numbers in lower right corners denote time elapsed from the first frame. Scale bars, 24 µm.

investigation of the signaling pathways mediated by these signals can elucidate mechanisms of OPC distribution. From our movies we also noticed that although OPCs often migrate in a directed fashion, the cells frequently change course.[12] Most interesting, is perhaps, the highly dynamic nature of OPC membrane processes, which are constantly extending and retracting from the onset of migration until axon wrapping, nearly 48 hours later. During this time, we observed that OPCs appear to interact with each other via their membrane extensions.[12] When neighboring OPCs have membrane processes that touch, each OPC then retracts its process and the two cells focus their extension activity in a different direction (Fig. 1c).[12] These findings raise the possibility that process activity and migration are influenced by contact-mediated signals between neighboring OPCs.

By 72 hpf, time-lapse imaging reveals that OPCs slow down their process activity and begin to wrap axons. Individual oligodendrocyte membrane extensions make contact with axon segments and small sheaths begin to form, and over time, elongate (Fig. 1d). By 96 hpf, axons are uniformly wrapped by a regular series of elongated sheaths.

One potential hypothesis to explain the regular spacing of oligodendrocytes along axons is that OPC process activity functions as a surveillance mechanism to recognize axons that are unassociated with OPCs. If this hypothesis is correct, then OPCs should divide and fill in areas of spinal cord where there are no neighboring OPCs. To test this hypothesis, OPCs within a 5-hemisegment region of the spinal cord labeled with the *Tg(olig2:egfp)* transgene were ablated at four days post-fertilization (dpf) with a pulsed-nitrogen laser. In the four days following ablation, OPCs in the ventral spinal cord and neighboring hemisegments divided and adjusted their position within the spinal cord by migrating into the ablated area.[12] Imaging *Tg(nkx2.2a:megfp)* larvae after ablation revealed that neighboring OPCs reorient their membrane processes toward the ablation site prior to migration.[12] These observations are consistent with the idea that OPCs migrate and divide to fill unoccupied space, which they assess using process activity, thereby ensuring uniform myelination.

Prior to these studies, the behaviors and interactions of OPCs within the spinal cord was poorly understood. Now, we are beginning to understand how OPCs migrate, interact and terminally differentiate into myelinating oligodendrocytes. Ultimately, we hope that these insights will aid in our interpretation of disease states and help us develop models of myelination disorders like Multiple Sclerosis (MS).

# Peripheral Nervous System

## Peripheral Glia Development

PNS axons, like CNS axons, must be myelinated by glia to ensure the rapid propagation of action potentials along their length. However, the mechanisms that mediate this ensheathment are poorly understood. Our lab is interested in utilizing the motor nerves of zebrafish embryos and larvae as models for studying the interactions between peripheral glia and their roles in motor nerve myelination.

Peripheral motor nerves are composed of motor axon-Schwann cell complexes that are ensheathed by flexible sheets of cells that are connected by tight junctions called the perineurium.[26-28] The perineurial sheath serves as a physical barrier that protects axons from ionic flux, toxins and infection. Within the perineurial sheath is the endoneurium. This nerve component is found outside of the axon-Schwann cell complexes, contains collagen fibers, fibroblast-like cells, blood vessels and macrophages.[28] Loosely encasing the nerve is the epineurium, which is composed of connective tissue, fibroblasts, mast cells, fat cells and lymphatic vessels.[29] Although these peripheral motor nerve components have been well described via ultrastructural analyses, very little is known about their origin or roles in development and maintenance of nerves.

The perineurium was first described in 1841, but for over 150 years, the origin of these cells, their interactions with neighboring cells and the roles they may play in motor nerve development have been poorly understood.[28] Previous studies in rodents have implicated interactions between Schwann cells and the perineurium during nerve development,[30] and although these studies implicate critical interactions, they still do not address the origin of perineurial cells.

One possible clue to the origin of perineurial cells comes from *Drosophila*, in which motor axon ensheathing glia originate from the lateral edges of the CNS and migrate out into the periphery via the CNS/PNS transition zone.[31-33] Because ensheathing glia in the fly serve a similar function to the vertebrate perineurium, that of a protective barrier, we hypothesized that the cells that make up the perineurium also arise within the CNS. To address this hypothesis as well as investigate the interactions between Schwann cells and the perineurium, we turned to time-lapse imaging in transgenic zebrafish embryos. As tools for further investigating these hypotheses, three transgenic lines have been very useful. The first, *Tg(sox10(7.2):mrfp)*, labels many neural crest derivatives including Schwann cells associated with motor nerves. The second, *Tg(nkx2.2a:megfp)*, labels ventral spinal cord p3 precursors and their derivatives and the third, *Tg(olig2:dsred2)*, labels motor neurons and their axons.[13]

To test the hypothesis that perineurial cells originate from the lateral edges of the spinal cord, we focused our attention on the *Tg(nkx2.2a:megfp)* transgenic line. Time-lapse imaging beginning at 48 hpf revealed that a subset of *nkx2.2a*[+] ventral spinal cord cells exited the CNS at iterated points along the anterior-posterior axis of the embryo, reminiscent of where motor axons exit (Fig. 2a).[13] Interestingly, *nkx2.2a*[+] membrane processes first exited at approximately 18 hpf, but it was not until 48 hpf that *nkx2.2a*[+] cell bodies were observed in the PNS. The molecular cues responsible for this behavior are not known, but we hypothesize that interactions with differentiating Schwann cells along the motor axons may hold the answer. To determine if perineurial glia exit the CNS at MEPs, we utilized time-lapse imaging in *Tg(nkx2.2a:megfp);Tg(olig2:dsred2)* embryos. These experiments revealed that *nkx2.2a*[+] cells only exit the CNS at MEPs (Fig. 2b), and once in the periphery, they divide and remain associated with the ventral surface of the spinal cord while dramatically extending membrane processes further distally along motor nerves (Figs. 2b and 2c).[13] By 72 hpf, they appear to ensheath motor axons and their associated Schwann cells. These data, along

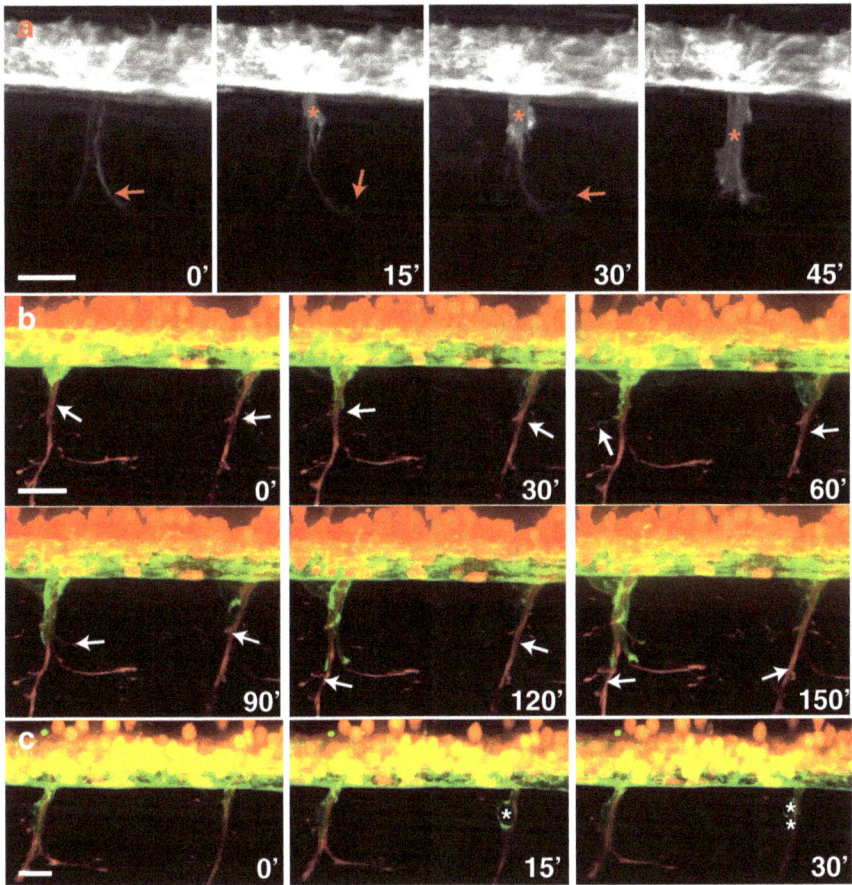

**Fig. 2** Behavior of peripheral glia revealed by *in vivo*, time-lapse imaging. (a) Frames captured from a 24-hour time-lapse sequence of a *Tg(nkx2.2a:megfp)* embryo beginning at 50 hpf. At 52 hpf (0-min time point), an EGFP+ process is observed exiting the spinal cord (red arrow) followed by the cell body (red asterisk), (b and c) Frames captured from a 24-hour time-lapse sequence of a *Tg(nkx2.2a:megfp);Tg(olig2:dsred2)* embryo beginning at 52 hpf. (b) At approximately 62 hpf (0-time point), peripheral EGFP+ processes (white arrows) are observed along DsRed+ motor axons. Over the next two hours, these processes extend like sheets further distally along the axons (white arrows), (c) At approximately 70 hpf (0-min time-point), an EGFP+ cell (white asterisk) in the PNS divides. Numbers in lower right corners denote time elapsed from the first frame. Scale bars, 24 μm.

with ultrastructural data, demonstrate that CNS-derived *nkx2.2a⁺* cells that associate with motor nerves subsequently form the perineurium.[13]

The close association of perineurial cells with motor axons and Schwann cells during development raised the possibility that perineurial cells influence motor nerve formation. To further test this, we combined live imaging with *nkx2.2a* loss-of-function experiments utilizing morpholino oligonucleotides (MO). These studies revealed that in the absence of perineurial cells, motor axons exit into the CNS via ectopic MEPs. Additionally, these axons were severely defasciculated and often wandered into neighboring muscle fields.[13] From these data we conclude that *nkx2.2a⁺* perineurial cells aid in motor axon guidance into the PNS via MEPs. Currently, we do not know if perineurial cells act as physical guideposts and directly interact with axons in the CNS to direct axons or if they express an attractant or repellant molecule that indirectly guides axons through the correct exit point.

To investigate potential interactions between perineurial cells and Schwann cells, we utilized *colourless (cls)* mutant embryos, which are deficient in Sox10, a transcription factor required for Schwann cell development.[34,35] In these embryos, Schwann cells associated with motor nerves die soon after 48 hpf, when perineurial cells normally exit the CNS. Visualization of perineurial cells via the *Tg(nkx2.2a:megfp)* transgene demonstrates that in *cls* mutants, perineurial cell bodies never exit the CNS, suggesting that they require direct interactions with differentiating Schwann cells to induce their migration and differentiation.[13] Interestingly however, perineurial cell processes extended into the periphery via MEPs. We hypothesize that perineurial cells in the CNS extend these processes to sense Schwann cell-derived cues. However, in the absence of Schwann cells, they fail to find it and their cell bodies never exit. Interestingly, motor axons exit the CNS in stereotypical positions in *cls* mutants, demonstrating that perineurial cells are present and still perform their initial role in nerve development by aiding in the guidance of motor axons into the PNS.[13]

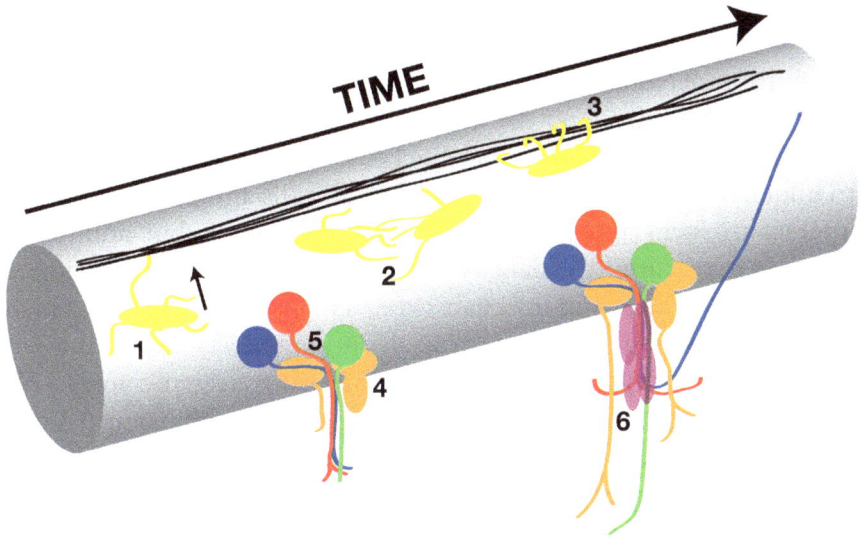

**Fig. 3** Schematic of glial behaviors in zebrafish embryos and larvae. *In vivo*, time-lapse imaging has revealed three main behaviors of OPCs: (1) OPCs migrate dynamically in the direction of their membrane processes, (2) OPCs appear to interact via their membrane processes before the initiation of axonal wraping and this behavior may direct the even distribution of OPCs in the spinal cord, and (3) OPC processes change morphology and can be observed ensheathing CNS axons. Imaging has also revealed three behaviors of peripheral glia and they include, (4) perineurial glia originate from the lateral edges of the spinal cord and exit via MEPs, (5) perineurial glia influence motor axon exit from the CNS, and (6) perineurial glia differentiation depends on interactions with Schwann cells along motor nerves. Yellow cells are OPCs, red, green and blue cells are the three primary motor neurons the zebrafish spinal cord. Orange cells are perineurial glia and purple cells are Schwann cells.

To determine if the interactions between Schwann cells and perineurial cells are reciprocal during motor nerve root organogenesis, mutant lines that have defective perineurial cell specification, migration or differentiation will be an important model for investigating these possibilities. Ultimately, by better understanding how individual components of the PNS interact, a clearer picture of the pathologies behind peripheral neuropathies and disease progression can be gleaned.

## Modeling PNS Disease

To fully understand the pathogenesis of peripheral neuropathies, we must first understand the molecular mechanisms that lead to the development of a fully functional nervous system. We are beginning to understand how motor axons, Schwann cells and perineurial cells interact to form a mature, myelinated nerve root, but additional research is required to identify the molecular players that mediate these orchestrated and intimate interactions. One way in which we are pursuing this is through mutagenesis screening of *Tg(nkx2.2a:mgfp)* and *Tg(sox10(7.2):mrfp)* transgenic lines. We are particularly interested in identifying mutants with perturbations in perineurial or Schwann cell number, distribution, morphology, viability or differentiation. From these, we anticipate identifying genes required for the interactions between PNS components. Ultimately, we hope that by understanding how the PNS is initially constructed, this may shed light on understanding how diseases alter it later in life.

## Conclusions

The CNS and PNS of the zebrafish are ideal models to investigate the developmental mechanisms required for neural and glial cell specification, migration and differentiation. Numerous transgenic lines specifically label many nervous system components and during embryogenesis and early larval development, the physical dimensions of the animals allow us to observe the behaviors of cells and how they interact with each other. By using time-lapse imaging we have been able to describe some of the developmental steps that ultimately result in myelinated axons in both the CNS and PNS. The visualization and characterization of the dynamic behavior and interactions of OPCs in the spinal cord as well as the communication between Schwann cells, motor axons and perineurial cells in the periphery would not have been possible without the coupling of *in vivo* imaging and genetics in the zebrafish (Fig. 3). Ultimately, we hope that the unique and powerful combination of these techniques will result in models of

CNS and PNS disease that will elucidate not only the causes, but progression of the diseases.

## References

1. Altmann CR and Brivanlou AH. Neural patterning in the vertebrate embryo. *Int Rev Cytol* 203:447–482, 2001.
2. Schmitz B, Papan C and Campos-Ortega JA. Neurulation in the anterior trunk region of the zebrafish *Brachydanio rerio*. *Roux's Arch Dev Biol* 202:250–259, 1993.
3. Papan C and Campos-Ortega JA. On the formation of the neural keel and neural tube in the zebrafish *Danio* (*Brachydanio rerio*). *Roux's Arch Dev Biol* 203:178–186, 1994.
4. Lewis KE and Eisen JS. From cells to circuits: development of the zebrafish spinal cord. *Prog Neurobiol* 69:419–449, 2003.
5. Beattie CE. Control of motor axon guidance in the zebrafish embryo. *Brain Res Bull* 53:489–500, 2000.
6. Eisen JS and Weston JA. Development of the neural crest in the zebrafish. *Dev Biol* 159:50–59, 1993.
7. Raible DW, Wood A, Hodsdon W, Henion PD, Weston JA and Eisen JS. Segregation and early dispersal of neural crest cells in the embryonic zebrafish. *Dev Dyn* 195:29–42, 1992.
8. Schneider VA and Granato M. Motor axon migration: a long way to go. *Dev Biol* 263:1–11, 2003.
9. Conchello JA and Lichtman JW. Optical sectioning microscopy. *Nat Methods* 2:920–931, 2005.
10. Graf R, Rietdorf J and Zimmermann T. Live cell spinning disk microscopy. *Adv Biochem Eng Biotechnol* 95:57–75, 2005.
11. Shin J, Park HC, Topczewska JM, Mawdsley DJ and Appel B. Neural cell fate analysis in zebrafish using olig2 BAC transgenics. *Methods Cell Sci* 25:7–14, 2003.
12. Kirby BB, Takada N, Latimer AJ, Shin J, Carney TJ, Kelsh RN and Appel B. *In vivo* time-lapse imaging shows dynamic oligodendrocyte progenitor behavior during zebrafish development. *Nat Neurosci* 9:1506–1511, 2006.
13. Kucenas S, Takada N, Park HC, Woodruff E, Broadie K and Appel B. CNS-derived glia ensheath peripheral nerves and mediate motor root development. *Nat Neurosci* 11:143–151, 2008.

14. Blader P, Plessy C and Strahle U. Multiple regulatory elements with spatially and temporally distinct activities control neurogenin1 expression in primary neurons of the zebrafish embryo. *Mech Dev* 120:211–218, 2003.

15. Kimura Y, Okamura Y and Higashijima S. Alx, a zebrafish homolog of Chx10, marks ipsilateral descending excitatory interneurons that participate in the regulation of spinal locomotor circuits. *J Neurosci* 26:5684–5697, 2006.

16. Sato T, Takahoko M and Okamoto H. HuC: Kaede, a useful tool to label neural morphologies in networks *in vivo*. *Genesis* 44:136–142, 2006.

17. Bernardos RL and Raymond PA. GFAP transgenic zebrafish. *Gene Expr Patterns* 6:1007–1013, 2006.

18. Higashijima S, Hotta Y and Okamoto H. Visualization of cranial motor neurons in live transgenic zebrafish expressing green fluorescent protein under the control of the islet-1 promoter/enhancer. *J Neurosci* 20:206–218, 2000.

19. Pfeiffer SE, Warrington AE and Bansal R. The oligodendrocyte and its many cellular processes. *Trends Cell Biol* 3:191–197, 1993.

20. Miller RH. Regulation of oligodendrocyte development in the vertebrate CNS. *Prog Neurobiol* 67:451–467, 2002.

21. Baumann N and Pham-Dinh D. Biology of oligodendrocyte and myelin in the mammalian central nervous system. *Physiol Rev* 81:871–927, 2001.

22. Park HC, Mehta A, Richardson JS and Appel B. Olig2 is required for zebrafish primary motor neuron and oligodendrocyte development. *Dev Biol* 248:356–368, 2002.

23. Zhou Q and Anderson DJ. The bHLH transcription factors OLIG2 and OLIG1 couple neuronal and glial subtype specification. *Cell* 109:61–73, 2002.

24. Calver AR, Hall AC, Yu WP, Walsh FS, Heath JK, Betsholtz C and Richardson WD. Oligodendrocyte population dynamics and the role of PDGF in vivo. *Neuron* 20:869–882, 1998.

25. McKinnon RD, Smith C, Behar T, Smith T and Dubois-Dalcq M. Distinct effects of bFGF and PDGF on oligodendrocyte progenitor cells. *Glia* 7:245–254, 1993.

26. Burkel WE. The histological fine structure of perineurium. *Anat Rec* 158:177–189, 1967.

27. Allt G. Ultrastructural features of the immature peripheral nerve. *J Anat* 105:283–293, 1969.
28. Bourne GH. *The Structure and Function of Nervous Tissue*. Academic Press, New York, 1968.
29. Olsson Y. Microenvironment of the peripheral nervous system under normal and pathological conditions. *Crit Rev Neurobiol* 5:265–311, 1990.
30. Parmantier E, Lynn B, Lawson D, Turmaine M, Namini SS, Chakrabarti L, McMahon AP, Jessen KR and Mirsky R. Schwann cell-derived desert hedgehog controls the development of peripheral nerve sheaths. *Neuron* 23:713–724, 1999.
31. Schmidt H, Rickert C, Bossing T, Vef O, Urban J and Technau GM. The embryonic central nervous system lineages of *Drosophila melanogaster*. II. Neuroblast lineages derived from the dorsal part of the neuroectoderm. *Dev Biol* 189:186–204, 1997.
32. Klambt C and Goodman CS. The diversity and pattern of glia during axon pathway formation in the Drosophila embryo. *Glia* 4:205–213, 1991.
33. Sepp KJ, Schulte J and Auld VJ. Peripheral glia direct axon guidance across the CNS/PNS transition zone. *Dev Biol* 238:47–63, 2001.
34. Dutton KA, Pauliny A, Lopes SS, Elworthy S, Carney TJ, Rauch J, Geisler R, Haffter P and Kelsh RN. Zebrafish colourless encodes sox10 and specifies non-ectomesenchymal neural crest fates. *Development* 128:4113–4125, 2001.
35. Kelsh RN and Eisen JS. The zebrafish colourless gene regulates development of non-ectomesenchymal neural crest derivatives. *Development* 127:515–525, 2000.

# Chapter 3

# Imaging the Cell Biology of Neuronal Migration in Zebrafish

Martin Distel, Jennifer Hocking and Reinhard W. Köster

*Helmholtz Zentrum München,*
*German Research Center for Environmental Health,*
*Institute of Developmental Genetics,*
*Munich-Neuherberg, Germany*

## Abstract

The migration of immature neurons from proliferation zones to their place of final differentiation occurs in many areas throughout the developing vertebrate central nervous system. As migration represents a key step in neuronal differentiation it has been modeled extensively in cell culture, matrix gels or explanted tissues. Due to the many complex interactions that are involved in this dynamic process, it would be ideal though to observe neuronal migration non-invasively directly in the developing organism. Furthermore, not only migratory pathways have to be characterized in detail, but also the underlying cell biology of neuronal migration needs to be revealed. For example, addressing how signal transduction is mediated into directed cellular movement at the level of different cellular organelles is not only crucial for understanding the cellular and molecular interplay of dynamic brain differentiation but also for revealing the etiology of neuronal migration diseases. Recent advances

---

Correspondence to: Dr. Reinhard W. Köster, Helmholtz Zentrum München, German Research Center for Environmental Health, Institute of Developmental Genetics, Ingolstädter Landstrasse 1, 85764 Munich–Neuherberg, Germany, E-mail: Reinhard.Koester@helmholtz-muenchen.de in zebrafish conditional genetics and simultaneous multi-cistron expression combined with the progress in high-resolution high-speed bio-optics promise that neuronal migration can be resolved at unprecedented detail in this organism. Thus the stage is set in zebrafish for true *in vivo* cell biology, which will allow for the re-addressing of the many models of neuronal migration derived from *in vitro* data of cultured cells or tissue explants. Hence zebrafish can serve to fuse the large fields of developmental genetics and cell biology in vertebrates.

*Keywords:* Zebrafish; Neuronal Migration; Cerebellum; Granule Cells; Bioimaging; Gal4; Cell Biology; Nucleokinesis; Centrosome.

# Introduction

Zebrafish embryos are ideal model organisms for the investigation of dynamic cell behaviors and their underlying molecular mechanisms in a vertebrate system. In particular, their external development, small size, fast embryogenesis and transparency mean that zebrafish can be used for continuous time-lapse imaging of *in vivo* cell movements. Combined with the steadily increasing repertoire of genetic methods, time-lapse studies have elucidated molecular processes regulating cell dynamics within the context of a living organism and such complexities as co-developing cell populations, simultaneously occurring tissue rearrangements or parallel acting signal transduction mechanisms. Thus imaging approaches in zebrafish help to verify and refine the models of cell behavior previously derived from *in vitro* studies of cultivated cells or explanted tissues (Fig. 1).

The dynamic process of neuronal migration is particularly complicated to mimic under culture conditions because of both the difficulty in cultivating neuronal progenitors of specific populations and the multiple players and interactions that are involved on the cellular and molecular level. Although the migratory pathways of neuronal progenitors and some of the molecular guidance mechanisms have been elucidated *in vivo*, the underlying cell biological processes that translate signal transduction events into directed cellular motility are only now beginning to be addressed. In this chapter we will propose that with the current knowledge about neuronal

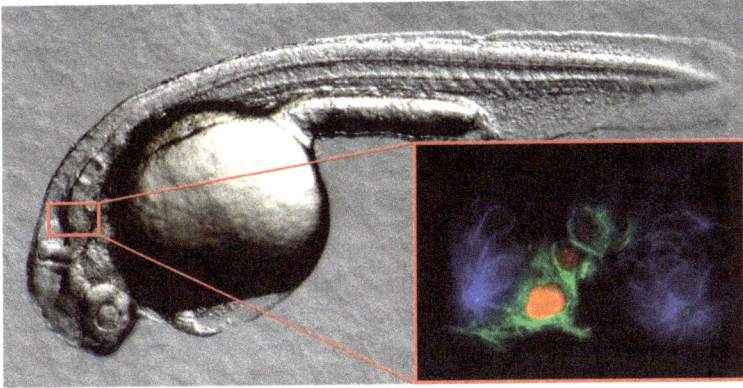

**Fig. 1** Zebrafish as a model organism for microscopic investigations. Transparent 24 hours post-fertilization (hpf) zebrafish embryo embedded in agarose. The inset shows subcellular structures in the living embryo, such as the cell nucleus in red, the microtubule cytoskeleton in green and cellular membranes in blue recorded using confocal fluorescence laser scanning microscopy.

migration mechanisms in zebrafish, the recent advances in bio-optics and genetic technologies as well as the possibilities for high resolution bio-imaging on both the spatial and temporal level, zebrafish could serve to fuse the two large research areas of developmental genetics and cell biology into a single research field of *in vivo* cell biology. Moreover, such an approach can be used to address the molecular mechanisms of subcellular dynamics regulating neuronal migration, in the context of organ differentiation and brain function. Thus in zebrafish embryos, bio-imaging at multiple levels will be able to inter-connect intracellular events with cell behavior, organogenesis, and eventually animal behavior in order to reveal a true development-structure-function relationship in a vertebrate model organism.

## Neuronal Migration

Embryonic development is characterized by extensive movements of tissues and cells. In some cases, cellular positions change passively through morphogenetic rearrangements. Alternatively, many cells undergo active migration away from their birthplace, in order to settle in a final position where their function is required in the mature organism. Some cell types, such a leukocytes or macrophages, are even destined to travel their entire life to fulfill control and safe-guarding functions.

The central nervous system is one area where well-ordered migra-tion of both neuronal progenitors and the growth cones of their protruding axons is of essential importance. This directed motility estab-lishes the highly ordered cellular composition of brain compartments and the correct wiring of the nervous system. In the vertebrate brain, migration of neuronal progenitors can be found in nearly every com-partment. For example, (a) the olfactory system in the anterior-most region of the forebrain shows pronounced neuronal migration from the subventricular zone, along the rostral migratory stream (RMS), and into the olfactory bulbs.[1] The RMS is of special importance as it is main-tained throughout life, providing a potential source of undifferentiated neuronal progenitors for therapeutic purposes.[2] (b) In the cortex, neu-ronal migration occurs mainly radially in a characteristic inward-out sequence, but it is accompanied by tangentially migrating cells emanat-ing from the medial and lateral ganglionic eminence.[3] (c) The development of the meso-diencephalic dopaminergic neurons requires extensive migration and it has been hypothesized that their positioning may contribute to differences in vulnerability to Parkinson's disease.[4] (d) In the cerebellum, granule neuron progenitors undergo a biphasic migration: tangential movement across the dorsal surface followed by a radial outward-in migration. Granule cell migration has been the focus of intense study to understand how neuronal migration is accomplished on the cellular and molecular level. However, other systems of interest in the posterior hindbrain include (e) the caudal migration of branchio-motoneurons and, (f) the circumferential migration of neurons forming the precerebellar system.[5] (g) Finally, radial migration occurs in the spinal cord along the entire length of the neural tube.

## Imaging Neuronal Migration in Zebrafish

Neuronal migration involves numerous and complex interactions among different cell populations and with the extracellular matrix, constraints from tissue boundaries, and coordination with concomitant tissue rearrangements. In addition, the migration of neuronal progenitors can occur over long distances and sometimes along complex routes.[6] In consideration of this natural complexity, processes of neuronal migration are ideally observed in the living organism and in a continuous manner involving methods of *in vivo* time-lapse microscopy. For such an approach the zebrafish embryo is ideally suited. Foremost, it is nearly transparent and small in size, allowing for excitation light to penetrate easily, while resulting fluorescence is not scattered extensively. In addition, the differentiation of the central nervous system occurs within two to three days; thus compared to other vertebrate model organisms the time-scale from neural proliferation, through migration, to terminal differentiation is condensed, allowing all of these events to be observed in a single time-lapse recording session.

The setup for time-lapse *in vivo* observations at cellular resolution is fairly simple as zebrafish can be mounted and anesthetized without major effort, oxygenation is not required and simple incubation chambers maintaining an ambient temperature of about 28°C can be constructed easily. Numerous detailed protocols for *in vivo* imaging of cellular behavior in zebrafish embryos have been established.[7–9] Due to their external development, zebrafish embryos can be easily manipulated to challenge migratory processes. Moreover, large collections of mutants are now available and many sophisticated genetic methods for tissue-specific and conditional genetics have been developed in recent years.[10–12] Finally, the collection of fluorescent proteins in many different colors, targeted to almost all subcellular locations and even being able to report many functional events inside cells, is expanding continuously.[13] Given the easy methods for generating transient and stable transgenics and the transparency of zebrafish embryos, this vertebrate model organism is ideal for investigating cellular and molecular processes of neuronal migration within a natural context.

## Neuronal Migration in the Zebrafish Cerebellum

Work in recent years from our and other labs has demonstrated that in particular the development of the zebrafish cerebellum is characterized by long-distance migration of neuronal progenitors.[6,14] These migratory cells arise from a dorsally positioned proliferation zone called the upper rhombic lip (URL). *In vivo* time-lapse studies indicated that neuronal progenitors migrate in an apparent two-phase migration: initially in an anterior direction towards the midbrain-hindbrain boundary (MHB), which delineates the anterior border of the cerebellum, and secondly, the progenitor cells turn at the MHB and migrate to more ventral positions (Fig. 2). Significant numbers of neuronal progenitors emanate from the URL for at least three days starting at about 28 hpf, but individual URL-derived cells pass their entire migratory route in about 1.5 to 2 days. Thus, *in vivo* time-lapse imaging in zebrafish is able to capture the entire migratory phase of individual cells, starting with cell divisions in the URL and ending with their terminal differentiation in ventral cerebellar regions.

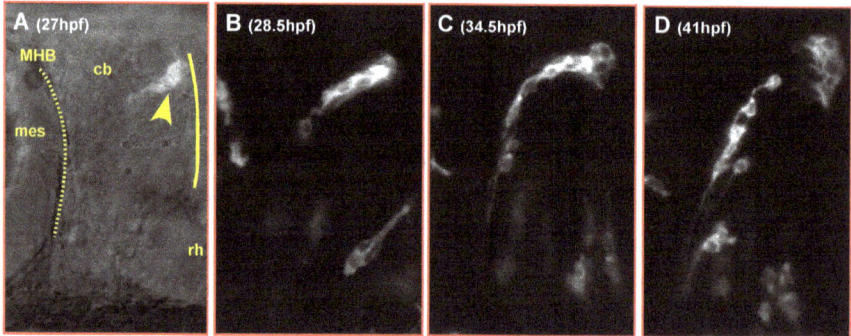

**Fig. 2** Migration of cerebellar neurons. (**A**) Lateral view on the brain of a living zebrafish embryo at 27 hpf. The midbrain-hindbrain boundary (MHB, dotted line) separates the mesencephalon (mes) from the rhombencephalon (rh). The cerebellum (cb) evolves from rhombomere 1 of the rhombencephalon. Neuronal precursors (arrow) labeled with GFP arise from the proliferative upper rhombic lip (yellow line). (**B**) Neuronal precursors start to migrate tangentially towards the MHB. (**C**) Precursor cells reaching the MHB turn ventrally in a second migratory step and start to extend first the axonal-like processes. (**D**) Neuronal precursors settle in ventral regions of rhombomere 1. All images are maximum projections of z-stacks taken from a time-lapse movie.

The URL does not give rise to the same cell type over the entire three or more days of generating migrating neuronal progenitors. First, during the second day of zebrafish embryogenesis until about 50 hpf, neuronal progenitors are generated for several brain stem nuclei positioned outside the cerebellum. These cells require for their motility the polysialylation of the neural adhesion molecule NCAM, as enzymatic removal of polysialic acid *in vivo* largely impaired their migration.[15] Subsequently, the URL begins to generate progenitors of the largest neuronal population in the vertebrate brain, the cerebellar granule neurons. Direct tracing of individual granule progenitor cells (GPCs), combined with retrograde axon labeling after terminal differentiation, indicated that a spatial pattern exists already inside the URL and reflects the different functional compartments of the mature cerebellum. While medial aspects of the URL generate GPCs that migrate in dorso-anterior directions and contribute to the non-vestibular cerebellar system, lateral URL regions provide GPCs that migrate along a latero-ventral route to generate granule neurons of the vestibular system of the cerebellum.[16] Recent studies combining time-lapse imaging of GPC migration with mutant analysis and molecular manipulation revealed that a different adhesion system consisting of classical Cadherins is playing a key role in orchestrating directional GPC migration, in particular of the dorsally migrating non-vestibular granule neurons.[17] Thus the cerebellar URL in zebrafish, like in other vertebrates, generates different neuronal cell populations over time, which employ different molecular adhesion systems to regulate their migration over long distances. Intriguingly, cerebellar development in zebrafish is very plastic and can be recapitulated during later stages. This has been demonstrated for early FGF signaling events emanating from the MHB, which are involved in patterning of the cerebellum,[18] but also for later stages of URL-derived GPC migration. Thus, when the entire primordium of the cerebellum is ablated during stages of prominent neuronal migration, FGF signaling from the MHB re-patterns the anterior hindbrain to re-establish cerebellar tissue. Subsequently, neuronal migration of GPCs becomes re-initiated, occurs along the characteristic antero-ventral pathway, and is followed by proper parallel fiber projection from regenerated granule neurons.[19]

This amazing capacity of the teleost cerebellum, which appears to last into adulthood,[20] provides a powerful system for the study of mechanisms of neuronal migration in the context of regenerating neuronal circuits in the vertebrate brain.

## Challenges of Imaging Neuronal Migration

The studies of motoneuron (see accompanying chapter by Stockinger and Heisenberg) and GPC migration, in the zebrafish hindbrain and cerebellum respectively, are but two examples that demonstrate the power of combining embryonic manipulation, molecular genetics and intravital time-lapse imaging in the unraveling of cellular behavior and its underlying molecular mechanisms in a natural context. However, as zebrafish is a fairly young model organism in the field of developmental genetics, many open questions remain. For example, the events triggering or halting migration have remained elusive so far, the guidance factors steering these neuronal populations along their pathways are mainly unidentified and the co-ordination of the migration of one neuronal population with the developmental behavior of co-developing neuronal populations is simply unknown. Answers to these questions will similarly come from studies of mutants or gene knock down, as well as by analyzing double transgenic zebrafish expressing differently colored fluorescent proteins in co-developing neuronal populations. For these investigations, the current technological level should be sufficient and technical advancements in the field of genetics or *in vivo* imaging are not necessarily required.

What needs to be addressed though to push the analysis of neuronal migration to the next level is to understand how molecular communication and signal transduction is translated into ordered subcellular and organelle behavior. Research over the past few decades in cell biology has contributed immense data sets addressing cellular motility and the regulation of subcellular dynamics. These data though are almost exclusively derived from cultured cells isolated from their natural environment and lacking the constraints and demands of brain compartment differentiation during embryogenesis. Thus models for neuronal migration derived from *in vitro* cell biology need to be transferred into and challenged in a novel field of *in vivo* cell biology. Given the fast embryogenesis, and genetic and imaging accessibility, the zebrafish is perfectly suited for merging the fields of developmental biology and cell biology in a vertebrate model organism. If intravital imaging can be pushed further in spatial and temporal resolution, then combined with advances in multi-transgene expression technologies for temporal and cell type specific control, the orchestration of neuronal migration will be understood in unprecedented detail. Thus, one day it may be possible by multi-level *in vivo* imaging in zebrafish to interconnect molecular processes with resulting organelle dynamics and cellular behaviors that in turn lead to proper organ differentiation and eventually control the behavior of the organism.

## Advances in Technology

### Confocal Laser Scanning Microscopy at High Speed

*In vivo* time-lapse imaging of neuronal migration at the cell biological level must be able to resolve the dynamics of subcellular components, which usually occur within the microsecond range. Thus, strong intrinsic contrast between cellular components and fast image recording technology at high sensitivity are each required. Recent advances in both biophysical optics and conditional zebrafish genetics mean that time-lapse *in vivo* cell biology of neuronal migration is now becoming possible.

Confocal microscopy is currently the method of choice for time-lapse imaging of living zebrafish at cellular resolution. However, the temporal resolution of image recording at full spatial resolution with conventional scan heads lies in the range of about 1 Hz. While this scan speed is more than sufficient to capture movements of entire cells, it is too slow to capture subcellular dynamics and behaviors such as fissions or fusions of mitochondria, cytoskeletal rearrangements or vesicle transport. Recently, various modifications to laser scanning confocal microscopy have increased the scan speed by one to two orders of a magnitude so that it now lies well within the range of cell biological events.

## Spinning Disk Confocal Microscopy

Spinning disk confocal microscopy uses a Nipkow disk for illumination and detection. This disk is equipped with a spiral pattern of holes and rotates through the excitation and emission light paths, thus converting the conventional single excitation beam into a parallelized multi-beam scanner (Fig. 3). Because the spinning disk is placed in a conjugate position to the focal plane of the objective, emitted fluorescence from planes above or below the focal plane will not pass

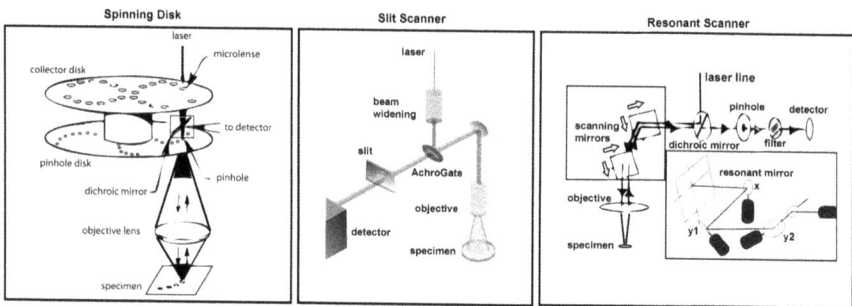

**Fig. 3** Confocal laser scanning at high speed. Spinning disk: Spinning disk confocal microscopes use a rotating Nipkow disk for image acquisition. The specimen is scanned with a multi-beam in a raster pattern using a fast horizontal scan (line scan) in conjunction with a slower vertical scan (frame scan) achieved by the spiral arrangement of holes in the collector disk. In recent microscopes the holes in the collector disk contain microlenses, which focus the laser beam through the pinholes in the pinhole disk for higher excitation efficiency and a better signal-to-noise ratio. The pinhole disk is also rotating, allowing only in focus information to pass through and reach the detector. Slit scanner: In contrast to point scanner microscopes slit scanner microscopes gain speed because they record entire lines instead of individual points. This is achieved by widening the laser beam to a line. The so-called AchroGate executes the function of the dichroic mirror of a point scan system to allow the excitation wavelength to reach the specimen, but only the emission wavelength to reach the detector. In order to obtain confocal images the pinhole in front of the detector is replaced by a slit in a slit scanning system. Resonant scanner: The difference between a resonant scanning and a traditional confocal point scanning system is the scanning mirror, which is oscillating at the so-called "Eigenfrequenz" at up to 8000 Hz. As a resonant scanning system uses regular pinholes, true confocal images can be obtained. The inset shows a resonant scanning system with two *y*-mirrors for precise illumination of the specimen and the *x*-mirror, which is oscillating at resonant frequency for fast scanning of a specimen.

through the center of the holes but will travel further sideways, eventually hitting the rim and reflecting away. This confocal effect allows one to image optical sections of a specimen, with the size of the holes in the Nipkow disk determining the thickness of the section. As the disk can rotate with up to several thousand rounds per minute, a raster scanning effect with low light exposure of the object is achieved, allowing for long-term time-lapse imaging of up to about 1000 frames per second at full frame size. Disadvantageous though, is that the pinhole cannot be variably adjusted in a continuous manner and that the system suffers from somewhat mediocre spatial resolution.[21,22]

## Slit-Scanning Confocal Microscopy

Instead of multiplexing, as performed by the spinning disk technology, slit scanners illuminate the entire line of a recorded image simultaneously by widening of the excitation laser beam (Fig. 3). Thus only unidirectional scanning is required for obtaining a full frame image, thereby reducing the scan time significantly. A slit that is aligned in a confocal manner in front of the detector assures optical sectioning and can be changed stepwise to project emitted fluorescence onto a line array of CCDs. At full frame size, this setup can reach image acquisition rates of about 100 Hz and is thus well suited for monitoring intracellular dynamics. As it is not a true point scanner, a reduction in spatial resolution along the slit axis, lack of precise optical manipulation (e.g. of photo-convertible fluorescent proteins) and discontinuous "pin-slit" adjustments are the trade-offs.[22]

## Resonant-Scanning Confocal Microscopy

Resonant scanners are a recent addition to the scan heads used in the field of confocal microscopy. Here, the principal of a traditional laser-scanning confocal microscopy system has remained almost the same. Point illumination with a pinhole in confocal arrangement is used for optical sectioning and the focused excitation beam is moved across the sample using moving mirrors (Fig. 3). If these mirrors are allowed to move at their "Eigenfrequenz" or resonant frequency, control over scan speed is lost but a significant increase in scan speed is gained. This resonant scanning does not reach the temporal resolution of a slit-scanning or spinning-disk system, but it still lies in the range of about 25 Hz at full frame resolution, which should be enough to resolve most cell biological processes. As with the slit scanner, faster frame rates can be achieved when the frame size is reduced, allowing a resonant scanner to operate above 100 Hz at small frame sizes. Of advantage is that the system it still a true point scanner with accompanying superior spatial resolution, the size of the pinhole can be continuously adjusted to allow for freely-defined optical manipulation, and multi-photon excitation can be combined with the speed of resonant scanning. Thus, nowadays several confocal laser-scanning technologies are available with sufficient spatio-temporal resolution for *in vivo* cell biology in living zebrafish embryos.

## Conditional Genetic Methods in Zebrafish

Monitoring neuronal migration often requires marking of the observed cells by extrinsic contrast. Since the discovery of the Green Fluorescent Protein, a whole battery of differently colored and functionalized fluorescent proteins has been identified or engineered.[13] Being genetically encoded, these vital dyes allow cell type-specific labeling to be achieved when combined with regulatory genetic elements. Therefore, the identification of cell type-specific enhancers for use in generating stable transgenic zebrafish strains with fluorescently-labeled cell populations has become one of the major areas of interest in the zebrafish research community. Such isolation of enhancers can be even more powerful when used for combinatorial genetic approaches. Here, rather than simply labeling a cell population of interest, the cell type-specific enhancer is used to drive the expression of a transcription factor or recombinase. This activator line, when crossed with specific responder strains, can not only be used to activate the expression of fluorescent reporters but also to mediate the expression of different transgenes. Since different activator and responder strains can be crossed together, such combinatorial genetic systems are multifunctional, save laborious efforts, make transgene expression data more comparable and allow for the establishment of embryonic-lethal disease models that are accessible for bio-imaging approaches. Mainly two different systems are currently being used.

### Cre-recombinase mediated expression

In the Cre system, a cell type-specific enhancer drives the expression of the bacteriophage P1-derived Cre-recombinase in the activator line. When crossed to a Cre-responder strain, this recombinase mediates the excision of DNA fragments that are flanked by sequence-specific recognition sites, called loxP sites. Commonly, such loxP-flanked spacer cassettes separate an enhancer from a transgene of interest, which only becomes expressed in the case of a tissue-specific recombination event. When expression in both the Cre-activator line and the Cre-responder line is under control of a cell type-specific enhancer, it is possible to even restrict transgene expression exclusively to the overlapping fraction of cells in which both enhancers are active. More sophisticated Cre-recombinase controlled genetic systems can be established if different sets of non-compatible loxP variants are used in combination and are arranged in a way to mediate a temporal sequence of Cre-recombination events, as has been demonstrated for the Flex system in mouse.[23] In zebrafish, a few reports show the usefulness of Cre-mediated transgene activation.[24] For example, expression of the murine *myc* oncogene under control of the zebrafish *rag2* enhancer causes lethality prior to sexual maturity due to aggressive tumor formation. Thus transgene-expressing carriers can only be maintained by *in vitro* fertilization. Separation of the *myc* oncogene cassette from the *rag2* promoter by insertion of a loxP-flanked spacer cassette allows for the propagation of transgenic carriers through subsequent generations, but tumor formation can be induced in a tissue-specific manner by injection of Cre-recombinase expression vectors into single-cell stage embryos.[25] Currently, the Cre-recombinase system is not widely used in zebrafish, probably because a large collection of tissue-specific Cre-activator lines is missing. However, with the recent advances in enhancer and gene trapping technologies in zebrafish, tissue-specific Cre-activator lines can be generated without major effort.

## Gal4-UAS system

Compared to Cre-recombinase, the Gal4 system, adopted from *Drosophila*, has recently obtained far more attention in zebrafish research, probably because experimental requirements have already been worked out in more detail. In this combinatorial system, an activator line drives the expression of the transcriptional activator Gal4, or a modified version of Gal4, under the control of a tissue-specific enhancer. Gal4-binding sites, called upstream activating sequences (UAS), mediate Gal4-dependent transgene expression in the effector line. When activator and effector strains are crossed together, tissue-specific transgene activation in the offspring is achieved by the tissue specificity of Gal4 expression (Fig. 4). Recently, transposon-mediated

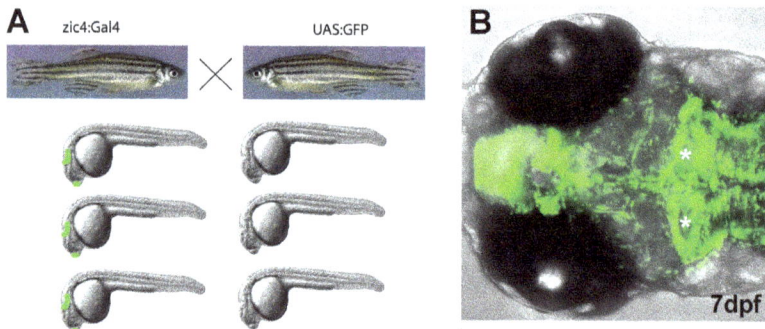

Fig. 4 Genetic techniques for manipulation of zebrafish. (**A**) A heterozygous Gal4 activator line (*zic4::Gal4*) is crossed to a homozygous effector fish (*UAS::GFP*). Neither of the parent fish expresses GFP, but half of the offspring express GFP in a tissue-specific manner. As the *zic4::Gal4* activator line carries the *Gal4* gene under control of the *zic4* enhancer, Gal4 is only expressed in distinct tissues and cell types such as cerebellar granule cells. Gal4 will transactivate the *GFP* transgene specifically in these cells so that cerebellar granule cells express the GFP protein. (**B**) Confocal image of a double transgenic embryo (*zic4::Gal4* and *UAS::GFP*) showing expression of GFP in the cerebellum (asterisks) and other *zic4* expression domains. Confocal image recorded at approximately seven days post-fertilization using a Zeiss LSM510 with a 20× objective.

enhancer and gene trapping have been used to generate libraries of many different tissue-specific Gal4-activator lines; thus for virtually any tissue of interest, a Gal4-activator line should now be available.[11,12,26,27]

To even restrict transgene activation to a smaller cell population that shares the co-expression of two genes of interest, the split-Gal4 technique can be employed.[28] Here, the DNA-binding domain of Gal4 and the transcriptional activation domain, both fused to heterodimerizing leucine zippers, are expressed under the control of separate cell type-specific enhancers. In cell types where both enhancers are active, overlapping expression is obtained and the DNA binding domain and the transactivation domain can reconstitute to form a functional Gal4 transcription factor. It requires though that two enhancers of interest are known and that triple-transgenic carriers are generated containing both Gal4 parts and a UAS-transgene cassette. Powerfully though, if one of the two Gal4 domains are expressed under control of a heat-shock inducible promoter, temporal as well as spatial control over transgene activation is achieved.

Recently, Gal4-combinatorial genetics and Cre-recombinase mediated transgene activation were combined in an elegant study to generate single transgene-expressing neurons with tissue-specific control in transient transgenic zebrafish. Here, tectal neuron-specific Gal4-mediated activation (*brn3a* promoter dependent) of a fluorescent reporter or transgene was restricted to a few cells by enabling Gal4 expression only in the case of a Cre-mediated activation event, which occurred at low efficiency because three different plasmids needed to be incorporated by a targeted cell of interest.[29]

In summary, these studies demonstrate that a number of genetic methods and their combinations are at hand or are currently being developed to achieve precise temporal and tissue-specific control of transgene expression in zebrafish. Thus, the controlled activation of cell biological reporters, ideally in only a few cells to create optimal contrast for use in time-lapse imaging approaches is within reach. Such precise spatio-temporal control of transgene expression is especially important for manipulating the cell biological events of neuronal migration in that the key players in cellular motility, polarity and adhesion are usually expressed ubiquitously and changes to their activity will cause severe and early phenotypes prior to nervous system development.

### Multi-transgene expression systems

While combinatorial genetics allows control over the where and when of transgene activation in specific cells, additional tools are necessary to address the cell biological events of neuronal migration. In particular, in order to understand their interdependence, simultaneous labeling and observation of different subcellular structures through the co-expression of multi-colored fluorescent markers is required. Moreover, addressing cell biological and molecular mechanisms of neuronal migration requires specific mechanism-manipulating genetic variants. This can be achieved through spatio-temporal control of transgene expression combined with multi-cistron expression systems — indeed making cell biological imaging of neuronal migration the gold standard for high-speed high-resolution single cell *in vivo* imaging (leaving open a platinum-standard for single molecule imaging *in vivo*).

### IRES expression vectors

Internal ribosomal entry sites (IRES) are probably the oldest method to achieve the co-expression of several transgenes from a single vector containing only one enhancer/promoter element. Here the ribosome can bind and start translation in a cap-independent mechanism that relies on internal secondary structures in the mRNA. Such poly-cistronic IRES-containing vectors have found widespread use in higher vertebrate model organisms such as mouse and chick. On the downside though, IRES-containing mRNAs mediate non-stoichio-metric expression of both cistrons, with the IRES-dependent cistron often being expressed at significantly reduced levels.[30] Furthermore, IRES sequences are very context dependent; thus their functionality is difficult to predict and has to be confirmed for every constructed vector. In zebrafish, IRES sequences have so far not received much attention. This is likely due to the low level expression obtained with the IRES sequences conventionally used in mouse and chick.[31,32] Low level expression however could be attributed to the fact that mostly mammalian IRES sequences have been tested so far, and the isolation of a zebrafish-inherent IRES sequence could probably revive this method. The reduced expression from IRES-dependent cistrons compared to cap-dependent cistrons could even be of use when different expression levels need be obtained. For example, labeling of some organelles like the centrosome with targeted fluorescent proteins quickly reaches saturation levels. This has the consequence that conventional cap-dependent expression often leads to the accumulation of excessive fluorescent protein in the cytoplasm, thereby significantly reducing the signal-to-noise ratio of centrosome labeling. In summary, the use of IRES-dependent expression in zebrafish should not be neglected completely, but instead be attempted again with IRES sequences isolated from teleosts (or from a teleost-infecting virus).

### 2A-peptides

In contrast to IRES sequences, viral 2A-peptides mediate strictly stoichiometric expression. In these self-processing polypetides, several cistrons are linked by small viral peptide sequences between 18–22 amino acids in length and usually ending with a PGP-sequence stretch. "Cleavage" between the penultimate glycine and the terminal proline residues does not occur post-translationally and does not involve additional co-factors. Rather peptide "cleavage" has been attributed to a sequence-dependent skip by the ribosome in forming a peptide bond.[33] Thus the production of proteins from two or more 2A-linked cistrons occurs at equimolar amounts and is highly efficient. Use of 2A-linked peptides was elegantly demonstrated in an approach to rescue CD3-deficient T-cells by expressing all four chains of the CD3-receptor complex in a 2A-peptide-linked arrangement in a single expression vector.[34] Given the global features of 2A-peptide "cleavage" characteristics, these peptides have been demonstrated to work in both animals and plants for obtaining reliable multi-transgene expression. Recently the technique has been successfully applied to zebrafish, in which virtually background-free 2A-peptide "cleavage" and subsequent proper sorting of differently targeted fluorescent proteins was demonstrated[35] (own unpublished data). It has to be kept in mind though that proteins translated from two 2A-linked cistrons carry parts of the 2A-peptide with them. The C-terminal protein is usually marked with a single proline residue at its N-terminus. Although this single amino acid may not have a significant influence on the protein's function, cDNAs of proteins with an N-terminal signal peptide should be positioned as the 5′ cistron. Unlike the almost unmodified C-terminal protein of a 2A-linked protein pair, the N-terminal protein is "tagged" by roughly 20 amino acids at its C-terminus. Thus it has to be verified that this C-terminal fusion does not interfere with the function of the protein. For example, peroxisome targeting, which is often mediated by the C-terminal-most amino acids of a protein, will be compromised if the peroxisomal protein is encoded by the 5′ cistron of a bicistronic 2A-linked expression construct. On the other hand this sequence tag could be turned into an advantage as antibodies have been successfully established against the 2A-peptide,[36] allowing for immunohistochemical detection, Western blot analysis and immuno-precipitation assays to be performed with the 5′ cistron-encoded protein. Given the robust co-expression of 2A-linked proteins in zebrafish, combined with the easy cloning procedure of simply adding 2A-sequences to a protein of interest by a PCR reaction, one can foresee that this powerful co-expression approach will obtain a lot of attention throughout the zebrafish research community. For bio-imaging approaches, several differently targeted fluorescent proteins could be used for simultaneously labeling a number of subcellular structures with unique colors, while for molecular studies transgene-expressing cells could be marked by the simultaneous 2A-mediated co-expression of a fluorescent reporter.

## Multicistronic Gal4 effectors

An alternative approach to obtain reliable co-expression of several transgenes, and one that completely avoids the generation of fusion proteins, is provided in conjunction with Gal4-activated transgene expression. As the transcriptional activator Gal4 is freely diffusible, it can bind to several independent UAS sequences and thereby activate a number of transgenes (Fig. 5). Furthermore, several UAS-dependent expression cassettes can be placed in a linear arrangement onto a single expression vector, which ensures co-segregation of several transgenes of interest, thereby allowing for the reliable fluorescent labeling of transgene-expressing cells.[29,37] Recent studies showed that two transgenes can also be activated from a single UAS sequence by flanking the UAS sites with two basal promoters in a mirror-image arrangement.[38] This "Janus" orientation facilitates cloning as it reduces recombination events in bacteria when two highly similar sequences are used in the expression vector (e.g. CFP/YFP). Moreover, co-activation of two transgenes does not require two independent binding events of Gal4 at different UAS sites of the expression vector, but theoretically a single binding event of Gal4 should be sufficient to activate transgene expression in two directions. This is supported by the finding that co-activation of two transgenes in a

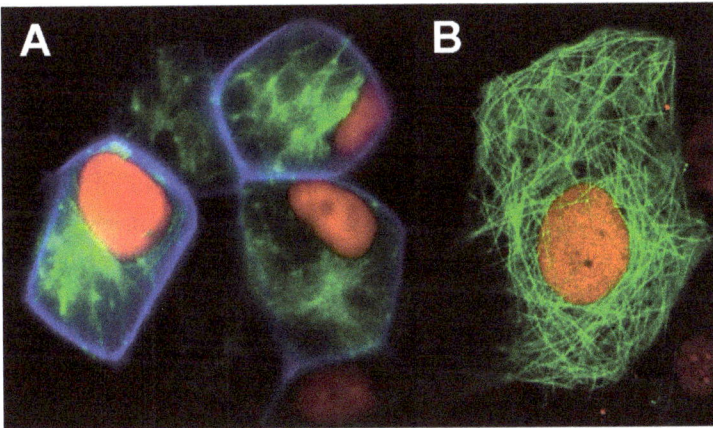

**Fig. 5** Subcellular labeling using multicistronic vectors. Using multicistronic UAS vectors several cellular components can be labeled simultaneously. (**A**) The nuclei of these cells of a living zebrafish embryo are labeled with RFP (red), microtubules are labeled with GFP (green) and the cellular membranes are labeled with CFP (blue). (**B**) Keratinocyte of a living zebrafish embryo with the nucleus labeled with RFP (red) and microtubules labeled with GFP (green). Images were recorded with a Zeiss LSM510 using a 40× water objective.

Janus vector can be achieved by a single UAS site (Distel and Köster, unpublished data). However, we cannot exclude that transgene activation involves cycles of Gal4 binding and release, with each new binding event involving the choice of selectively activating only one transgene at a time. Nevertheless, co-activation of transgene expression occurs reliably, although not in equimolar amounts. Quantification of transgene expression showed that usually one transgene is expressed in slightly higher amounts up to about 30% more. One advantage of this multi-cistronic UAS arrangement is that different expression modules can be prepared in advance, allowing for very efficient cassette-like cloning procedures to generate new co-expression constructs. In addition, several of these expression cassettes can be combined on a single vector, allowing four or more different transgenes to be co-expressed. Moreover, as pointed out above, strong transgene expression is not always of advantage as it can reduce signal-to-noise ratio in the case of saturating fluorescent reporter protein expression. Further, high expression levels can lead to significant overexpression phenotypes, such as those often observed with the expression of microtubule-binding fluorescent reporter proteins. Multi-cistronic UAS cassette vectors allow one to account for these problems as expression levels can be adjusted through varying the number of Gal4-binding sites in the UAS sequence stretch. While strong expression can be reached with five UAS sites in tandem order, reducing the number of UAS sites in the sequence stretch successively reduces the expression level of the transgene. For example, strong nuclear and membrane labeling can be achieved with a 5× UAS Janus cassette, while moderate but sufficient expression levels can be obtained from the same vector from a 2× UAS centrosome/microtubule-labeling Janus unit (unpublished observations). This possibility to fine-tune expression levels of transgenes is particularly valuable for addressing cell biological questions of neuronal migration, and for monitoring and influencing migratory processes without interfering with other basic cellular processes such as cell proliferation.

In summary, several valuable methods for the efficient and reliable co-expression of numerous transgenes have been established in recent years. For more sophisticated approaches these methods can of course be combined, for example to drive 2A-peptide linked transgene expression from several UAS cassettes. In addition, cell type-specific combinatorial expression systems have been established for zebrafish and a number of cell type-specific enhancers or expression-activating transgenic strains are readily available. Further, the recent advances in bio-optics now allow for the observation of subcellular dynamics at sufficient spatial and temporal resolution without significant photo-bleaching or photo-toxicity. Thus the stage is set for true *in vivo* cell biology in zebrafish, a vertebrate model organism that will allow for the challenging of the many models for neuronal migration derived from *in vitro* data of cultured cells or tissue explants. Hence zebrafish can serve to fuse the large fields of developmental genetics and cell biology in vertebrates. Given that currently significant progress is being made in the areas of single molecule imaging and in the automatic monitoring of zebrafish behavior, imaging at many organismic levels is coming into sight in zebrafish. In summary, tracing individual molecules, observing subcellular dynamics and monitoring cell behavior can be connected to the processes of organ differentiation and ultimately explain consequences for animal behavior, providing a truly dynamic structure-function relationship.

## Cell Biology of Neuronal Migration: More Questions Than Answers

Expression analysis, fate mapping approaches and time-lapse studies have revealed many migratory pathways of neuronal populations in the vertebrate brain. Moreover, pharmacological manipulations, conditional mutagenesis and analysis of genetically mosaic animals have provided insight into how cellular motility, directionality, coherence and pathfinding of neuronal migration are regulated at a molecular level. But how are these molecular signals translated into coordinated subcellular dynamics to achieve directional motility? What are the subcellular consequences of interactions between migratory and non-migratory cells and how do groups of cells jointly orchestrate their cellular migration machineries?

## Coordination of Nucleokinesis

First insights into these complex mechanisms were derived from analyzing the normal functions of molecules altered in neuronal migration diseases, collectively called lissencephalies. Lissencephaly-causing genes (when compromised in their function), such as *lis1, nde, ndel* and *dcx,* are highly conserved from slime molds to humans. Interestingly, the proteins that they encode are usually associated with the microtubule cytoskeleton and/or the centrosome. When these cellular structures were analyzed in detail during neuronal migration, it became evident that many neuronal progenitors do not translocate as an entire body with all their cellular components (a migratory mechanism termed somal translocation), but rather that their organelles are transported forward in a well-choreographed sequential manner. Initially, the neuronal progenitor forms and extends a leading process, which is subsequently filled by microtubule fibers. Most characteristic for this migratory mode is that the centrosome travels in front of the nucleus, based on the direction of travel, and then translocates into the leading process. This centrosomal movement extends the centrosomal-nuclear distance and is believed to trigger subsequent nuclear forward movement towards the centrosome. Finally, the cytoplasm travels behind (Fig. 6). Forward movements of neuronal progenitors alternate with resting phases, giving the appearance of saltatory motion. The orchestration of this coordinated set of organelle movements it thought to be mediated by the microtubule skeleton, which is organized by the centrosome through its function as a microtubule organizing center (MTOC). Microtubules connect the leading process with the centrosome and from there surround the nucleus in a characteristic perinuclear microtubule cage that is believed to mediate the forward transport of the nucleus during nucleokinesis.[39–41] The protein products of lissencephaly type I-causing genes are often involved in microtubule-nucleus coupling and so it is thought that the nucleus in affected neurons of these patients is unable to follow the migration-initiating movements of the centrosome. As a consequence, neuronal progenitors stall close to or inside of proliferation zones and fail to reach their proper place of function.[40]

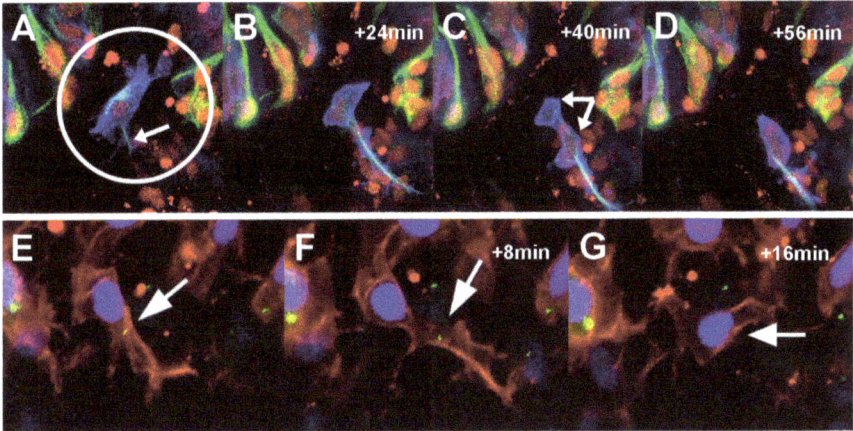

**Fig. 6** Visualization of subcellular dynamics during cell migration. Using a multi-cistron subcellular labeling technique, several cellular components were labeled in migrating cells undergoing nucleokinesis in a living zebrafish embryo at approximately 28 hpf. (A–D) The nucleus of the cell (white circle in A) is labeled with RFP (red), the cell membrane with CFP (blue) and the microtubules with GFP (green). Images are taken from a time-lapse movie recorded with a Zeiss LSM510 using a 40× water objective. (**A**) The cell starts to establish a process (arrow) in the later direction of migration. A bundle of microtubules (green) is already visible in this process. (**B**) Microtubules extend more and more and the process elongates. (**C**) The nucleus (red) is translocated in the direction of migration (two arrows showing the translocation), leaving a cytoplasm-filled membrane behind. The movement of the nucleus occurs in a saltatory fashion. (**D**) The trailing membrane is retracted, hereby finishing the cycle of nucleokinesis. (E–G) The nucleus of this cell of interest (white circle in E) is labeled with CFP (blue), the membrane with RFP (red) and the centrosome with YFP (yellow). The migratory step is initiated by the movement of the centrosome. From **E** to **F** the distance between nucleus and centrosome (arrow) increases, indicating the forward movement of the centrosome. (**G**) The nucleus is translocated in the direction of the centrosome movement. The centrosome disappears as it moves out of the focal plane. Images are taken from a time-lapse movie recorded with a Zeiss LSM510 using a 40× water objective.

The current understanding of nucleokinesis though is rudimentary and partially contradictory results have been published. In addition, data from isolated, cultured neuronal progenitors, cells grown in collagen, matrigel or similar assays, tissue explants and organotypic slice cultures and using different neuronal populations are often compared to one another. Furthermore, although both radially and tangentially migrating neuronal progenitors have been shown to migrate via nucleokinesis, homotypic and heterotypic migratory modes clearly involve different cell-cell interactions, suggesting that variations of a common theme of nucleokinesis have to be taken into account. This is a field where zebrafish can contribute significantly as an *in vivo* cell biology model without the need to extensively manipulate the cells that are to be observed.

One current open question in understanding nucleokinetic migration regards the driving force behind movement of the nucleus. Studies on the function of lissencephaly-causing genes suggest that the nucleus follows the centrosome in a Dynein-dependent manner along the microtubule fibers that interconnect the centrosome with the nucleus. Pharmacological inhibition of the actin-myosin network though suggested that instead of a microtubule-mediated "pulling" force, non-muscle myosin II-mediated pushing from the rear of the cell is responsible for nuclear movements towards the centrosome.[42] Which of these forces is active *in vivo* remains to be clarified and first requires a careful analysis of the coordination of the microtubule as well as the actomyosin system in relation to the different steps during neuronal progenitor forward movement. Specific *in vivo* perturbations of both of these systems in several migrating neuronal populations need to be performed in order to reveal whether nuclear forward movement relies on one force and whether this machinery is reused in neuronal populations that migrate in different modes. URL-derived neuronal progenitors and branchiomotor neurons in the zebrafish cerebellum and hindbrain respectively, may represent interesting model cell types for answering these questions because the hindbrain is highly accessible for pharmacological treatments through injections of the fourth ventricle. Rather than acting in an either/or mechanism, both systems, microtubule-dependent nuclear "pulling" and actomyosin-dependent nuclear "pushing," may well act together. The question that arises then is how do both systems coordinate and synchronize their activity mechanistically and on the molecular level. Do they regulate different aspects of nuclear forward transport or do they cooperate as partners in the same process? Detailed *in vivo* time-lapse studies at high temporal resolution in zebrafish embryos could help to resolve these questions and thus provide further insights into the etiology of lissencephaly.

## The Centrosome as Maître de Danse?

Given its characteristic positioning and its prominent role in inter-connecting the leading process with the nucleus through organizing the microtubule network, the centrosome has achieved major atten-tion in the field of neuronal migration research. Yet its role remains undefined and evidence for its direct involvement in organizing nucleokinesis is largely circumstantial. In fact, in a recent time-lapse study it was demonstrated that in radially migrating cerebellar gran-ule progenitor cells, the nucleus can overtake the centrosome during nucleokinetic forward movement and that perinuclear microtubules do not converge at the centrosome.[43] Nevertheless, during resting periods of the cell, the centrosome was found in front of the nucleus and intereference with Lis1 demonstrated an uncoupling of centro-some dynamics with nuclear movements. Thus rather than serving as a permanent guide for the nucleus, the centrosome could play a preparatory role during cell resting and leading process elongation in setting up the next forward nuclear movement. Whether such a role for the centrosome holds true for tangentially migrating neurons, in which the centrosome has been reported to stay strictly in front of the nucleus, remains to be shown and probably requires a similar high-resolution *in vivo* study to unravel the temporal order of organelle dynamics.

Prior to neuronal migration, the centrosome plays an important role in neuroblast division and the maintenance of apico-basal polar-ity. Neural cell divisions usually occur with the mitotic spindle aligning parallel to, and the division axis orthogonal to, the ventricu-lar zone. After such a symmetric division, the centrosome moves to the apical cell surface while the nucleus resumes so-called interkinetic nuclear movement until it returns to the ventricular zone for the next round of mitosis. Until now the molecular and cell biological mechanisms that ensure repositioning of the centrosome at the apical membrane, and thus re-establishment of apico-basal polarity in both daughter cells, remain largely elusive. Similarly, the driving forces of interkinetic nuclear movements, which are thought to occur to provide space at the ventricle for cell divisions, are mostly unknown. How is the centrosome transported and anchored to the apical ventricular side? Does the centrosome indeed re-establish apico-basal polarity *de novo* after every round of neuroblast mitosis or do other cues maintain polarity during cell divisions, with the centrosome simply following these cues in resuming its position?

Although there is still debate about the existence of clearly asymmetric neuroblast divisions responsible for the generation of the neuronal precursors that emigrate from germinal zones, initiation of neuronal migration must involve significant rearrangements in the centrosome-nuclear relationship. Nucleokinesis would require that the centrosome leaves the apical membrane and repositions itself in front of the nucleus; furthermore centrosome-nucleus coupling needs to be established, as these organelles do not coordinate their movements during interkinetic nuclear migration. Thus, one could define the onset of migration by following the repositioning of the centrosome *in vivo*. Indeed, first results in the zebrafish cerebellum suggest that interkinetic nuclear movements can occur over significant spatial distances, thereby obscuring the onset of real neuronal migration and the interpretation of molecular guidance mechanisms (own unpublished results).

Another important question that arises with respect to centrosome function during neuronal migration is its role in initiating and determining axon formation. Elegant studies have shown that explanted neuronal progenitors rely on intrinsic mechanisms to establish neuronal polarity, with the centrosome selecting the nearest neurite to become the future axon.[44] Interestingly, time-lapse imaging has shown that *in vivo* axon formation and projection can occur simultaneously to neuronal migration, with the leading process extending to become the forming axon.[14] This is in good agreement with the centrosome traveling close to and moving into the leading process during nucleokinetic neuronal migration. However, elegant *in vivo* time-lapse studies demonstrated for retinal ganglion cells in the developing zebrafish eye that the position of the centrosome does not predict the site of axon formation *in vivo*; in fact, axon formation was found to occur basally after the cells have left the apical membrane, but prior to centrosome positioning close to the nucleus.[45] Whether these findings are a common scheme of axon formation by migrating neurons or are confined to specific migratory modes needs to be addressed.

Finally, centrosome amplification is often found in cancerous cells and may be responsible for the genetic instability of tumorigenic cells. Interestingly though, a recent study in Drosophila demonstrated that flies with extra centrosomes in about 60% of their cells develop normally and maintain a stable diploid genome over several generations.[46] During most cell divisions, spindle formation is achieved by centrosome clustering; however the asymmetric division of neuroblasts, which usually precedes and initiates neuronal migration, is perturbed and results in symmetric divisions that give rise to more proliferating cells. In addition, transplanted brain cells with multiple centrosomes can induce metastatic tumors in host flies. This finding suggests that neuroblastoma formation could result from a failure in proper centrosome-mediated neuronal migration, keeping neural progenitors in contact with proliferation signals inside germinal zones for too long. Time-lapse imaging of the behavior of neuroblasts with induced extra centrosomes in tissues of pronounced neuronal migration in zebrafish would reveal if neuronal migration can be accomplished by centrosome clustering or whether migration is stalled but leaves the neuroblasts competent for proliferation.

## Organelle Dynamics During Neuronal Migration

Clearly the centrosome is not the only organelle that needs to contribute to neuronal migration. While it can be postulated that cellular organelles are simply dragged along, cell biological observations argue against such a passive role. For example, along with the centrosome, the Golgi apparatus and the endoplasmic reticulum (ER) are found in front of the nucleus, with at least the Golgi preceding nuclear movement during nucleokinesis.[42] Thus the first question that arises is whether the centrosome precedes and controls Golgi movements or whether these organelles move independently from one another. Is there a temporal order in which specific organelles are moved inside migrating neuronal progenitors and do they use the same molecular mechanisms? Are Lis1-defective cells also impaired in Golgi or ER transport? What are the molecular and cellular mechanisms that trigger organelle forward movement? One could postulate that forward extension of the leading process shifts the center of gravity to the cell's front and thus organelles become readjusted in their position along microtubule fibers with respect to that center. In such a model, adhesion factors in connection with the cytoskeletal networks must play an important role in transmitting tension. Intriguingly, recent studies about molecular tracing revealed that adhesion factors are far more flexible than assumed and are transported inside cells and along the membrane.[17,47,48] They could thus play an active role in shifting tension forces inside migrating neuronal progenitors and relocating the center of gravity, which is then followed by organelle repositioning. To answer these types of questions *in vivo* would require even further improvement in the temporal and spatial resolution of bio-imaging techniques in zebrafish and the development of intelligent fluorescent reporter systems. Thus the next challenge for zebrafish imaging is already in sight.

## Conclusions

Neuronal migration is a key step in nervous system development and in the integration of adult neural stem cells into functional circuits. Although migratory pathways have been delineated, the molecular and cell biological mechanisms underlying neuronal migration are just beginning to be understood. In recent years, research in zebrafish has primed this model organism to address the *in vivo* mechanisms of neuronal migration. Major pathways of migrating neurons have been characterized in several regions of the nervous system and significant advances in optical imaging, combinatorial genetics and multi-transgene expression systems have been achieved. Thus the stage is set in zebrafish to merge the large research fields of cell biology and developmental genetics in order to decipher the subcellular coordination of directed cellular motility and its molecular regulatory mechanisms. Many exciting open questions await this research field and the superb bio-imaging properties as well as the genetic accessibility of zebrafish will allow us to watch how cells organize their internal machineries to accomplish one of the most difficult tasks in central nervous system differentiation. This *in vivo* cell biology will challenge many existing models of cellular and neuronal migration to verify, modify or redefine our knowledge of the events that occur within a moving cell in the context of a living and developing organism. Zebrafish researchers can look forward to a bright future.

## Acknowledgments

We wish to apologize to all colleagues whose work could not be cited due to limited space. We thank all members of our research group for intense discussions. M. Distel is supported by the Studienstiftung des deutschen Volkes, J. Hocking is a fellow of the Natural Sciences and Engineering Research Council of Canada and R.W. Köster is financed by a BioFuture Grant Award (0311889) of the German Ministry of Education and Research (BMBF), a grant of the German Research Foundation (DFG KO1949/3-1) and the European Commission Coordination Action ENINET (LSHM-CT-2005–19063).

## References

1. Murase S and Horwitz AF. Directions in cell migration along the rostral migratory stream: the pathway for migration in the brain. *Curr Top Dev Bio* 61:135–152, 2004.
2. Soares S and Sotelo C. Adult neural stem cells from the mouse subventricular zone are limited in migratory ability compared to progenitor cells of similar origin. *Neuroscience* 128:807–817, 2004.
3. Nadarajah B and Parnavelas JG. Modes of neuronal migration in the developing cerebral cortex. *Nat Rev Neurosci* 3:423–432, 2002.

4.  Jacobs FM, Smits SM, Hornmann KJ, Burbach JP and Smidt MP. Strategies to unravel molecular codes essential for the development of meso-diencephalic dopaminergic neurons. *J Physiol* 575:397–402, 2006.
5.  Bloch-Gallego E, Causeret F, Ezan F, Backer S and Hidalgo-Sanchez M. Development of precerebellar nuclei: instructive factors and intracellular mediators in neuronal migration, survival and axon pathfinding. *Brain Res Rev* 49:253–266, 2005.
6.  Mione M, Baldessari D, Deflorian G, Nappo G and Santoriello C. How neuronal migration contributes to the morphogenesis of the CNS: insights from the zebrafish. *Dev Neurosci* 30:65–81, 2008.
7.  Cooper MS, D'Amico LA and Henry CA. Analyzing morphogenetic cell behaviors in vitally stained zebrafish embryos. *Methods Mol Biol* 122:185–204, 1999.
8.  Langenberg T, Brand M and Cooper MS. Imaging brain development and organogenesis in zebrafish using immobilized embryonic explants. *Dev Dyn* 228:464–474, 2003.
9.  Distel M and Köster RW. *In vivo* time-lapse imaging of zebrafish embryonic development. *Cold Spring Harb Protoc* 2:doi:10.1011/pdb.prot4816, 2007.
10. Hardy ME, Ross LV and Chien CB. Focal gene misexpression in zebrafish embryos induced by local heat shock using a modified soldering iron. *Dev Dyn* 236:3071–3076, 2007.
11. Scott EK, Mason L, Arrenberg AB, Ziv L, Grosse NJ, Xiao T, Chi NC, Asakawa K, Kawakami K and Baier H. Targeting neural circuitry in zebrafish using GAL4 enhancer trapping. *Nat Methods* 4:323–326, 2007.
12. Asakawa K, Suster ML, Mizusawa K, Nagayoshi S, Kotani T, Urasaki A, Kishimoto Y, Hibi M and Kawakami K. Genetic dissection of neural circuits by Tol2 transposon-mediated Gal4 gene and enhancer trapping in zebrafish. *Proc Natl Acad Sci USA* 105:1255–1260, 2008.
13. Miyawaki A. Innovations in the imaging of brain functinos using fluorescent proteins. *Neuron* 48:189–199, 2005.
14. Köster RW and Fraser SE. Direct imaging of *in vivo* neuronal migration in the developing cerebellum. *Curr Biol* 11:1858–1863, 2001.
15. Rieger S, Volkmann K and Köster RW. Polysialyltransferase expression is linked to neuronal migration in the developing and adult zebrafish. *Dev Dyn* 237:276–285, 2008.
16. Volkmann K, Rieger S, Babaryka A and Köster RW. The cerebellar rhombic lip is spatially subdivided in producing granule cell populations of different functional compartments. *Dev Biol* 313:167–180, 2008.

17. Rieger S, Senghaas N, Walch A and Köster RW. Cadherin-2 controls directional chain migration of cerebellar granule neurons. *PLoS Biol* 7:e1000240, 2009, doi 10.1371/journal.pbio.1000240.

18. Jaszai J, Reifers F, Picker A, Langenberg T and Brand M. Isthmus-to-midbrain transformation in the absence of midbrain-hindbrain organizer activity. *Development* 130:6611–6623, 2003.

19. Köster RW and Fraser SE. FGF signaling mediates regeneration of the differentiating cerebellum through repatterning of the anterior hindbrain and reinitiation of neuronal migration. *J Neurosci* 26:7293–7304, 2006.

20. Zupanc GKH. Adult neurogenesis and neuronal regeneration in the central nervous system of teleost fish. *Brain Beha Evol* 58:250–275, 2001.

21. Nakano A. Spinning-disk confocal microscopy — a cutting-edge tool for imaging of membrane traffic. *Cell Struct Funct* 27:349–355, 2002.

22. Conchello J-A and Lichtman JW. Optical sectioning microscopy. *Nat Methods* 2:920–931, 2005.

23. Floss T and Schnütgen F. Conditional gene trapping using the FLEx system. *Methods Mol Biol* 435:127–138, 2008.

24. Hans S, Kaslin J, Freudenreich D and Brand M. Temporally-controlled site-specific recombination in zebrafish. *PLoS ONE* 4:e4640, 2009.

25. Langenau DM, Feng H, Berghmans S, Kanki JP, Kutok JL and Look AT. Cre/lox-regulated transgenic zebrafish model with conditional myc-induced T cell acute lymphoblastic leukemia. *Proc Natl Acad Sci USA* 102:6068–6073, 2005.

26. Davison JM, Akitake CM, Goll MG, Rhee JM, Gosse N, Baier H, Halpern ME, Leach SD and Parsons MJ. Transactivation from Gal4VP16 transgenic insertions for tissue-specific cell labeling and ablation in zebrafish. *Dev Biol* 304:811–824, 2007.

27. Distel M, Wullimann MF and Köster RW. Optimized Gal4 genetics for permanent gene expression mapping in zebrafish. *Proc Natl Acad Sci USA* 106:13365–13370, 2009.

28. Luan H, Peabody NC, Vinson CR and White BH. Refined spatial manipulation of neuronal function by combinatorial restriction of transgene expression. *Neuron* 52:425–436, 2006.

29. Sato T, Hamaoka T, Aizawa H, Hosoya T and Okamoto H. Genetic sinlge-cell mosaic analysis implicates ephrinB2 reverse signaling in projections from the posterior tectum to the hindbrain in zebrafish. *J Neurosci* 27:5271–5279, 2007.

30. Mizuguchi H, Xu Z, Ishii-Watabe A, Uchida E and Hayakawa T. IRES-dependent second gene expression is significant lower than

cap-dependent first gene expression in a bicistronic vector. *Mol Ther* 1:376–382, 2000.

31. Köster R, Götz R, Altschmied J, Sendtner R and Schartl M. Comparison of monocistronic and bicistronic constructs for neurotrophin transgene and reporter gene expression in fish cells. *Mol Marine Biol Biotechnol* 5:1–8, 1996.

32. Fahrenkrug SC, Clark KJ, Dahlquist MO and Hackett PB. Dicistronic gene expression in developing zebrafish. *Marine Biotechnol* 1:552–561, 1999.

33. Donnelly ML, Luke G, Mehrotra A, Li X, Hughes LE, Gani D and Ryan MD. Analysis of the aphthovirus 2A/2B polyprotein 'cleavage' mechanism indicates not a proteolytic reaction, but a novel translational effect: a putative ribosomal 'skip'. *J Gen Viro* 82:1027–1041, 2001.

34. Szymczak AL, Workman CJ, Wang Y, Vignali KM, Dilioglou S, Vanin EF and Vignali DA. Correction of multi-gene deficiency *in vivo* using a single 'self-cleaving' 2A petide-based retroviral vector. *Nat Biotechnol* 22:589–594, 2004.

35. Provost E, Rhee J and Leach SD. Viral 2A peptides allow expression of multiple proteins from a single ORF in transgenic zebrafish embryos. *Genesis* 45:625–629, 2007.

36. Ryan MD and Drew J. Foot-and-mouth disease virus 2A oligopeptide mediated cleavage of an artifical polyprotein. *EMBO J* 13:928–933, 1994.

37. Köster RW and Fraser SE. Tracing transgene expression in living zebrafish embryos. *Dev Biol* 233:329–346, 2001.

38. Paquet D, Bhat R, Sydow A, Mandelkow E-M, Berg S, Hellberg S, Fälting J, Distel M, Köster RW, Schmid B, *et al.* A zebrafish model of tauopathy allows *in vivo* imaging of neuronal cell death and drug evaluation. *J Clin Invest* 119:1382–1395, 2009.

39. Marin O, Valdeolmillos M and Moya F. Neurons in motion: same principles for different shapes? *Trends Neurosci* 29:655–661, 2006.

40. Kerjan G and Gleeson JG. Genetic mechanisms underlying abnormal neuronal migration in classical lissencephaly. *Trends Genet* 23:623–630, 2007.

41. Higginbotham HR and Gleeson JG. The centrosome in neuronal development. *Trends Neurosci* 30:276–283, 2007.

42. Bellion A, Baudoin JP, Alvarez C, Bornens M and Metin C. Nucleokinesis in tangentially migrating neurons comprise two alternating phases: forward migration of the Golgi/centrosome associated with centrosome splitting and myosin contraction at the rear. *J Neurosci* 25: 5691–5699, 2005.

43. Umeshima H, Hirano T and Kengaku M. Microtubule-based nuclear movement occurs independently of centrosome positioning in migrating neurons. *Proc Natl Acad Sci USA* 104:16182–16187, 2007.

44. de Anda FC, Pollarolo G, Da Silva JS, Camoletto PG, Feiguin F and Dotti CG. Centrosome localization determines neuronal polarity. *Nature* 436:704–708, 2005.

45. Zolessi FR, Poggi L, Wilkinson CJ, Chien CB and Harris WA. Polarization and orientation of retinal ganglion cells *in vivo*. *Neural De* 1:2, 2006.

46. Basto R, Brunk K, Vinadogrova T, Peel N, Franz A, Khodjakov A and Raff JW. Centrosome amplification can initiate tumorigenesis in flies. *Cell* 133:1032–1042, 2008.

47. Kametani Y and Takeichi M. Basal-to-apical cadherin flow at cell junctions. *Nat Cell Biol* 9:92–98, 2007.

48. Ogata S, Morokuma J, Hayata T, Kolle G, Niehrs C, Ueno N and Cho KWY. TGF-beta signaling-mediated morphogenesis: modulation of cell adhesion via cadherin endocytosis. *Genes Dev* 21:1817–1831, 2007.

Chapter 4

# Applications of Fluorescence Correlation Spectroscopy in Living Zebrafish Embryos

Xianke Shi,* Yong Hwee Foo,* Vladimir Korzh,†
Sohail Ahmed,‡ and Thorsten Wohland*

*Department of Chemistry, National University of Singapore
† Institute of Molecular and Cell Biology, Singapore
‡Institute of Medical Biology, Singapore

## ABSTRACT

Recent advances in light microscopy and spectroscopy make it possible to study single molecule behavior in an intact 3D multicellular organism. Taking advantage of the transparent tissues of zebrafish embryos as well as the established molecular and genetic approaches in this animal model, we present in this chapter some examples of studying molecular dynamics and interactions in living zebrafish embryos by fluorescence correlation spectroscopy (FCS) and single wavelength fluorescence cross-correlation spectroscopy (SW-FCCS). FCS/SW-FCCS are ultra-sensitive experimental techniques that analyze fluorescence fluctuations from a static observation volume and provide information about the concentration, diffusion constant and biophysical properties of the fluorescent particle. A short introduction of the theory and instrumentations of FCS/FCCS is provided. We then give an overview of new FCS modalities that are designed to cope with emerging challenges in intracellular applications, e.g. correction of cell membrane movement during measurement and correction of spectral cross-talk of fluorescent proteins.

Correspondence to: Dr. Thorsten Wohland, Department of Chemistry, National University of Singapore, 3 Science Drive 3, Singapore 117543, E-mail: chmwt@nus.edu.sg

A short description of the preparation of zebrafish embryos for FCS measurements is included. In the last section we demonstrate measurements of blood flow velocities with high spatial resolution, measurements of diffusion coefficients of fluorescently labeled cytosolic and membrane-bound proteins within living zebrafish embryos, and the determination of dissociation constants of interacting proteins by SW-FCCS. The extension of this biophysical tool into living organism allows answering developmental biology questions directly on a molecular basis, and provides a platform to address the potential artifacts arising from Petri dish-based *in vitro* studies.

*Keywords*: Fluorescence Correlation Spectroscopy; Zebrafish; Single Molecule Events; Blood Flow Velocity; Diffusion Coefficient; Dissociation Constant.

# Introduction

The quantitative determination of molecular interactions has become an important subject in biology. Especially systems biology, which aims at understanding a complex biological system like the cell on the basis of its molecular building blocks and their interactions, urgently requires input from biophysical techniques with quantitative numbers for on- and off-rates, enzymatic and reaction rates and affinities of biomolecular interactions. Until recently, single molecule sensitive measurements, let alone quantitative measurements, were restricted to *in vitro* samples or at best to cell cultures. Several developments in the last decades were necessary to finally bring fluorescence spectroscopy into live animals. We will list just some of the key developments. First came the detection of a single fluorescent molecule in solution by Hirschfeld.[1] This demonstrated that single molecule detection is possible and started a surge in new fluorescence spectroscopy methods which still lasts today,[2,3] and includes fluorescence correlation spectroscopy (first developed by Madge *et al.*[4] and recently reviewed by Krichevsky and Bonnet[5]), the method discussed in this chapter. The second important development was the discovery and application of genetically encoded fluorophores,[6-8] for which the 2008 Nobel Prize in Chemistry was given. This allowed tagging molecules in a specific fashion with a defined stoichiometry, avoiding many of the problems posed by external dyes. And although fluorescent proteins come with their own set of problems (duration of maturation, complicated fluorescence behavior, photo-stability to name a few), research in the field provides researchers with a wide choice of fluorophores with improved properties.[9,10] The third development was the introduction of *Danio rerio*, zebrafish, as a model system for developmental biology. As a vertebrate it is evolutionary closer to humans than *Drosophila melanogaster*, the working horse in developmental biology, but it has a much shorter reproduction cycle than mice, another widely used model animal. But its main advantage from the point of optical spectroscopy is its small size and its transparency, allowing penetration and measurements of almost the whole embryo by simple one-photon excitation.

Fluorescence Correlation Spectroscopy (FCS) is a technique which requires single molecule sensitivity. By measuring fluorescence signal fluctuations stemming from single molecules, FCS is capable to obtain the parameters characterizing the underlying molecular dynamics. In particular, FCS obtains accurate (relative) concentrations as well as diffusion coefficients. The number of secondary parameters which have been derived from the concentration and diffusion coefficients include among others chemical reactions,[11] dissociation constants and stoichiometry of binding,[12-14] aggregation,[15] protein membrane interactions,[16] lipid membrane characteristics,[17] protein and lipid raft association,[18] transient binding times,[19] and GPCR trafficking behavior.[20] In literature one can find a rising number of reviews of FCS,[5,21-23] its biological applications,[24] and comparison with other methods.[25] However, only recently was FCS introduced for *in vivo* studies of complex 3D tissues, breaking the last barrier to the use of physiologically relevant, quantitative, and single molecule sensitive measurements within living organisms. These new developments will be described in more detail later in this chapter. First, we will start with a description of FCS and its principles, and subsequently, introduce a range of new modalities, e.g. fluorescence cross-correlation spectroscopy (FCCS), which made recent progress in FCS in organisms possible.[26] Then, we will give a short overview of measurements performed already in zebrafish by this method and describe measurements performed in other animals, including *Drosophila melanogaster* and *Caenorhabditis elegans*, before ending the chapter with a short summary and outlook.

# Introduction to FCS

FCS is a fluorescence-based ultrasensitive experimental technique, developed to measure molecular dynamics and interactions on the single molecule level. It possesses single molecule sensitivity, but at the same time, it is based on fast statistical treatment of recorded data. Conventional FCS records and analyzes the fluorescence intensity from a small observation volume through which fluorescently labeled particles can freely move. It is important in FCS that the signal-to-noise ratio is sufficient to be able to distinguish fluorescence fluctuations caused by single molecules from the noise and background level. The seemingly noisy fluorescence intensity fluctuations recorded from a fast photon-counting detector contain the raw information about the various processes that the fluorescent molecules undergo when passing through the observation volume. The resulting fluorescence intensity fluctuations can be analyzed by autocorrelation to obtain information about the molecular processes causing these variations. Parameters such as local concentrations, diffusion coefficients and photo-physical dynamics of the fluorescent particles can be extracted (Fig. 2). As the name suggests, autocorrelation analysis correlates the fluorescence signal at time point $t$, with itself at a different time point $t + \tau$, in which $\tau$ is referred to as the lag time or time of correlation. This can give an insight into the time course of the underlying process causing the fluorescence fluctuations, as the correlation can only be found if the underlying process persists longer than the time $\tau$. The normalized autocorrelation function (ACF) is defined as:

$$G(\tau) = \frac{\langle F(t) \cdot F(t + \tau) \rangle}{\langle F(t) \rangle^2}$$

where $\langle ... \rangle$ denotes the time average and $F(t)$ is the fluorescence intensity at time $t$. Autocorrelation analysis can be performed using hardware or software correlators,[27-29] and the experimental ACF curves are then mathematically fitted with the appropriate model of the system dynamics to obtain corresponding parameters. In FCS it is important to know the theoretical model of the process under investigation to be able to properly fit the data. In the most common case, the diffusion of particles is observed, for which the appropriate fit model for FCS is well known (see below). Any influence on the diffusion coefficient by changes of local environment of a particle or by binding can then be monitored and evaluated.

The standard FCS setup nowadays uses a confocal illumination scheme (Fig. 1), as for instance described by Ricka and Binkert in 1989[30] or by Rigler *et al.* in 1993.[31] Previous FCS measurements suffered from a poor signal-to-noise ratio due to the technical difficulty to create a small enough observation volume. A small observation volume ensures that a minimum number of molecules are detected, so the fluctuations caused by single molecules can be distinguished and can contribute significantly to the autocorrelation function. In general the signal-to-noise ratio of FCS is improved when a smaller number of molecules are seen since the size of the fluctuations of single molecules compared to the average intensity is increased.[32] In the confocal illumination scheme, the use of a higher numerical aperture objective and a small pinhole in the setup generate an observation volume less than one femtoliter (~0.5 fL). When the concentration of the molecule-of-interest drops to ~3 nanomolar (nM), there is, on average, only one molecule at any one time in the observation volume. In general, FCS can measure samples with about 0.1–1000 molecules in the observation volume, corresponding to the sub-nanomolar to the micromolar concentration range. Assuming the observation volume created in the confocal setup has a Gaussian intensity profile in all three directions, the ACF for a single component translational diffusion is given by:[33]

$$G(\tau) = \frac{1}{N}\left(1 + \frac{\tau}{\tau_d}\right)^{-1}\left[1 + \left(\frac{\omega_0}{\omega_z}\right)^2\frac{\tau}{\tau_d}\right]^{-\frac{1}{2}} + G_\infty$$

with $\tau_d = \dfrac{\omega_0^2}{4D}$

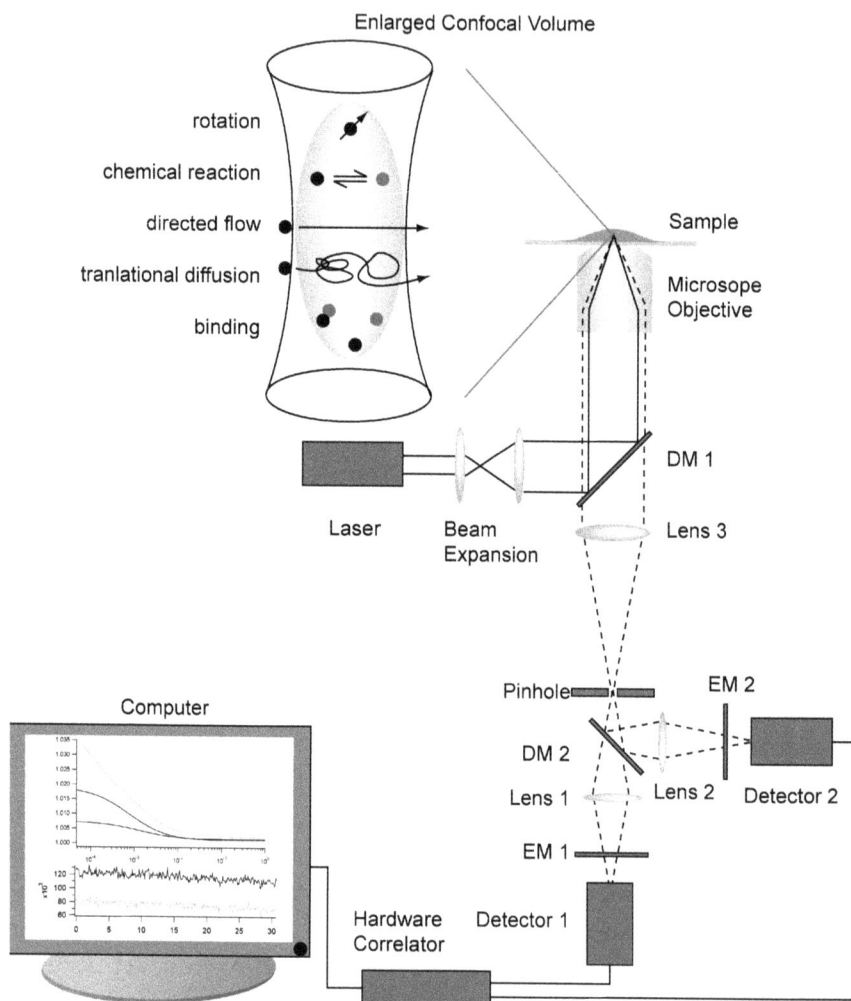

**Fig. 1** A typical confocal FCS setup. This dual-color single wavelength fluorescence cross-correlation spectroscopy (SW-FCCS) setup utilizes one laser beam with one photon excitation (OPE) to excite two fluorophores. The laser beam is expanded first to fill completely the back aperture of the microscope objective. The objective then focuses the beam into the sample. The emitted fluorescence light is separated from excitation light by a dichroic mirror and then focused by a tube lens. Out-of-focus light is spatially filtered by a pinhole at the conjugate plane, which restricts the observation volume. A second dichroic mirror is placed after the pinhole to separate the signals into two channels. The separated fluorescence signals then pass through corresponding emission filters and are finally focused onto two avalanche photodiode

where $N$ is the apparent number of particles in the observation volume; $\tau_d$ is the diffusion time required for one molecule to diffuse through the observation volume; $\omega_0$ and $\omega_z$ are the radial and axial extends of the observation volume; $G_\infty$ is the convergence value of the ACF for long times and is usually very close to its theoretical value of 1; and $D$ is the diffusion coefficient of the fluorescent molecule. For a more extensive list of example correlation functions see Ref. 34.

The introduction of the confocal illumination scheme increased the popularity of FCS in the following years and the field grew at an accelerating pace. Recently, a number of new technologies, e.g. stimulated emission depletion (STED)[19,35] and technologies using near-field phenomena,[36,37] have been introduced to further reduce the observation volume. This promises an even higher sensitivity and enables FCS measurement in more concentrated samples.

FCS is well suited for intracellular applications. They are non-invasive in nature and highly sensitive. The spatial resolution of FCS is defined by the size of the observation volume, usually less than one micrometer in dimension and can be further reduced to the nanometer scale. In typical biological samples, concentrations of biological molecules range from 1 nM to 1 µM, which results in about 1–1000 particles in the observation volume, a range just measurable by FCS. Therefore, FCS can directly be used to study protein dynamics at their physiological expression levels. At the same time, FCS provides a wide range of temporal information from microseconds to seconds. This allows the measurement of a broad spectrum of biological molecules, whose diffusion behavior can be extremely diverse in living biological samples. Furthermore, new instrumentations that combine FCS with imaging techniques, e.g. confocal laser scanning microscopy, also expand the

---

**Fig. 1** (*Continued*) detectors (APDs) that count the incoming photons and send a TTL pulse for each photon to the hardware correlator. The correlator calculates the auto- and cross-correlation online in a semi-logarithmic time scale that is displayed on a computer. The auto- and cross-correlation functions reveal processes that cause the fluorescence fluctuations in the observation volume, i.e. rotational diffusion, chemical reaction, directed flow, translational diffusion, and molecular bindings. DM: Dichroic mirror; EM: emission filter.

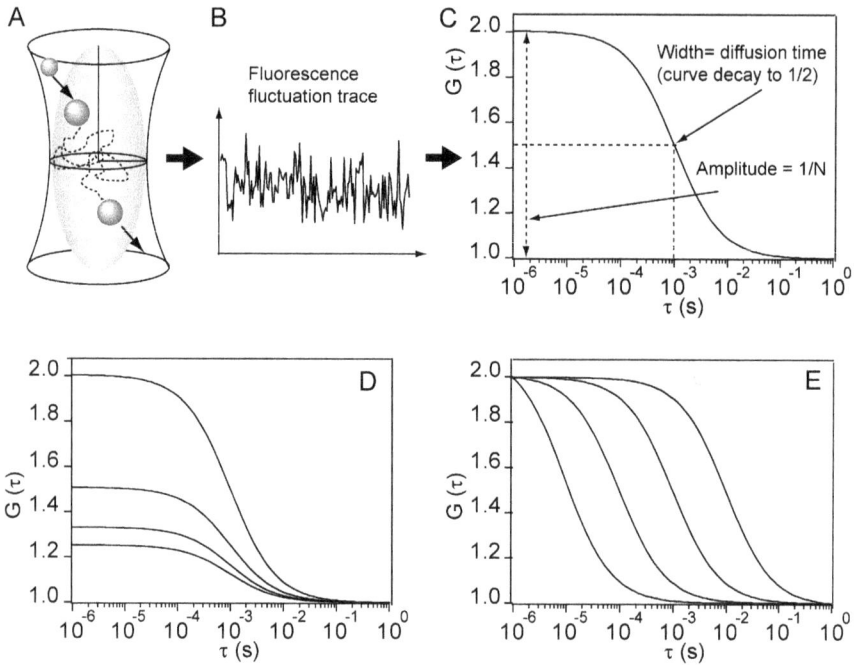

**Fig. 2** Characteristics of autocorrelation functions (ACFs). **(A)** Fluorescent particles diffuse through an observation volume, which give rise to **(B)** fluorescence intensity fluctuations. (C) Autocorrelation analysis generate ACF curves which give information about the concentration (number of particles in the observation volume) and diffusion coefficient (diffusion time required for each molecule to diffuse through the observation volume) of this molecule. The amplitude of the ACF curve is inversely proportional to the number of particles, and the width of the ACF curve gives the diffusion time. **(C)** Examples of ACFs of different sample concentrations. Number of particles from top to bottom: 1, 2, 3, 4, corresponding to concentrations (assuming 0.5 fL of observation volume): 3.3, 6.6, 10.0, 13.3 nM. **(D)** Examples of ACF of different diffusion coefficients. Diffusion time from left to right: 0.01, 0.1, 1, 10 ms, corresponding to diffusion coefficients (assuming the observation volume has a Gaussian intensity profile in all three dimensions): $1.7 \times 10^{-5}$, $1.7 \times 10^{-6}$, $1.7 \times 10^{-7}$, $1.7 \times 10^{-8}$ cm$^2$/s.

applicability of FCS in biological samples. The combination, also known as fluorescence correlation microscopy (FCM),[38] allows the user to obtain an image of the sample first, before identifying a position on the image where subsequent FCS measurements can be performed.

This is especially useful in intracellular applications. The typical volume of a eukaryotic cell is $10^{-12}$ L, which is three orders of magnitude larger than the observation volume of FCS. Using FCM, protein dynamics thereby can be specifically investigated in subcellular compartments.

Recent advances in the methodology of FCS have also led to the development of numerous FCS variants, including fluorescence cross-correlation spectroscopy (FCCS), two-photon excitation FCCS (TPE-FCCS), single wavelength FCCS (SW-FCCS), pulsed inter-leaved excitation FCCS (PIE-FCCS), two-focus FCCS, and scanning FCS, which will be briefly discussed in this section.

Dual-color FCCS utilizes two-color labeling to specifically study molecular interactions. Previously, single-color FCS was used to measure molecular binding, e.g. receptor-ligand interactions,[12,14,39] based on the theory that relative changes in mass upon binding lead to a reduction in the diffusion coefficient. The efficiency of single-color FCS to resolve binding is limited, as in order to distinguish two components (free and bound particles) in FCS, their diffusion coefficients must differ by at least a factor of 1.6,[40] which, according the Stokes-Einstein relation, requires a mass change of at least a factor of 4. In addition, single-color FCS cannot resolve specific binding between individual components of a multi-component system. Therefore, the concept of multi-color FCCS was introduced by Eigen and Rigler in 1994[41] and the experimental real-ization was achieved by Schwille *et al.* in 1997.[42] In dual-color FCCS, both binding partners of interest are labeled with distinct fluorescent tags. The two labels are simultaneously excited and fluorescence signals are collected in separated spectral detection channels using dichroic mir-rors and filters. Aside from the autocorrelation of signals from each channel, the signals from both channels are cross-correlated. The nor-malized cross-correlation function (CCF) is written as:

$$G(\tau) = \frac{\langle F_1(t) \cdot F_2(t+\tau) \rangle}{\langle F_1(t) \rangle \langle F_2(t) \rangle}$$

where $F_1(t)$ and $F_2(t)$ represent the fluorescence intensities at time $t$ in two different channels. Binding between the two molecules leads to concurrent movement of the two labels, which in turn produces a positive signal in the cross-correlation analysis. FCCS determines molecular binding based on the relative amplitude of the cross-correlation curve to the autocorrelation curves, and thus is independent of molecular mass and provides a basis for quantitative analysis irrespective of the relative size of the binding partners. In addition, FCCS is independent of distance between and orientation of the fluorophores, and it therefore represents an attractive alternative to fluorescence resonance energy transfer (FRET) measurements which are typically used to study molecular interactions.[25] The potential of FCCS to effectively measure molecular interactions was demonstrated both *in vitro*[43–46] and *in vivo*.[47–50]

The first experimental realization of FCCS utilized two lasers to excite two different fluorophores.[42] This setup achieves highest molecular brightness, which is also known as counts rate per molecule per second (*cps*), as laser wavelengths can be chosen accordingly to match the absorption maximum of each fluorophore, and fluorophores can also be selected to have widely separated emission maxima to minimize cross-talk. However, the realization of dual laser excitation requires the alignment and maintenance of a perfect overlap of the two observation volumes. This is complicated by chromatic aberrations and differences of the focal spot size resulting from different laser wavelengths. The complication is further worsened in biological samples, as living cells are optically in-homogeneous. It has been shown that misalignment of the two laser beams in FCCS measurements will cause reduced cross-correlation amplitudes and a shift of CCF curve towards slower decays.[51] Thus a quantitative determination of biomolecular interactions using dual laser excitation approach can be difficult in living cells, let alone in a multicellular living organism. An obvious solution to this problem is to use a single light source instead of two to simultaneously excite two fluorophores. This was first achieved using two photon excitation (TPE).[52] TPE describes a nonlinear process in which two photons are absorbed within a very short time interval ($<10^{-15}$ s) by one fluorophore. The molecule is promoted to an excited state with the combing energy of the two photons and then follows the normal fluorescence emission pathway.[53] This nonlinear process is theoretically symmetry-forbidden and exhibits different selection rules. Consequently, the TPE excitation spectra of common fluorophores are considerably different from that of one photon excitation (OPE), although the emission spectra remain the same. In fact, the TPE excitation spectra of different fluorescent dyes exhibit large overlaps,[54] providing the basis for simultaneous excitation of two fluorophores with a single light source. The experimental realization of TPE-FCCS was first reported by Heinze *et al.* in 2000[55] using a pulsed Ti-Sapphire laser. As one light source was used, system alignment is greatly simplified and instrument stability is dramatically improved.

Another advantage of TPE is its intrinsically restricted observation volume, without the use of a pinhole. In order to achieve the simultaneous absorption of two photons by one fluorophore ($<10^{-15}$ s) high photon flux densities are required since the probability of TPE is proportional to the square of the illumination intensity (the probability of OPE is directly proportional to the illumination intensity). Thus when the TPE laser beam is focused on the sample by a microscope objective, only the center of the focus has a high enough photon density to achieve TPE. The size of the intrinsically restricted observation volume is comparable to that of OPE confocal microscopy. This makes TPE especially advantageous when applied in multicellular organisms, as multiple cell layers cause strong and multiple light scattering. In OPE confocal microscopy, which achieves optical sectioning with a detection pinhole that reject out-of-focus light, the fluorescence light of interest can be scattered and partially blocked by the pinhole itself. Thus TPE with a non-descanned detection scheme [56] can achieve a higher signal-to-noise ratio and a deeper working distance in organisms.

One drawback of TPE is the low molecular brightness achieved. This is believed to be caused by high laser power induced fluorophore photo-bleaching [57,58] and saturation.[59] The use of fluorescent proteins (FPs) in biological samples further worsens the situation, as FPs are not as photostable as chemical dyes and can be more easily photo-bleached. Previous studies have shown that the signal-to-noise ratio of FCS is determined by the molecular brightness of the fluorophore.[60] Thus the signal-to-noise ratio of TPE-FCCS can be relatively poor. In addition, the high cost of the pulsed Ti-Sapphire laser system reduced the availability of TPE-FCCS. These led others to develop new FCCS modalities. OPE single wavelength FCCS (SW-FCCS) was introduced by Hwang and Wohland.[61] The principle of SW-FCCS borrowed the concept of TPE-FCCS, except OPE is used instead of TPE (see Fig. 1 for the SW-FCCS optical setup). Without the largely overlapping excitation spectra of fluorophores by OPE, the experiment is only feasible with fluorophore pairs that possess similar excitation maxima but largely different Stokes shifts. In Hwang and Wohland's report, the use of fluorescein and tandem dyes/Quantum Dots as fluorophore pair proved feasible. In the following years, this technique was established to measure protein-protein interactions in living cells[62,63] and even in living embryos[26] with the FP pair of enhanced green fluorescent protein (EGFP) and monomeric red fluorescent protein (mRFP) (see Fig. 3 for their spectra). New chemical

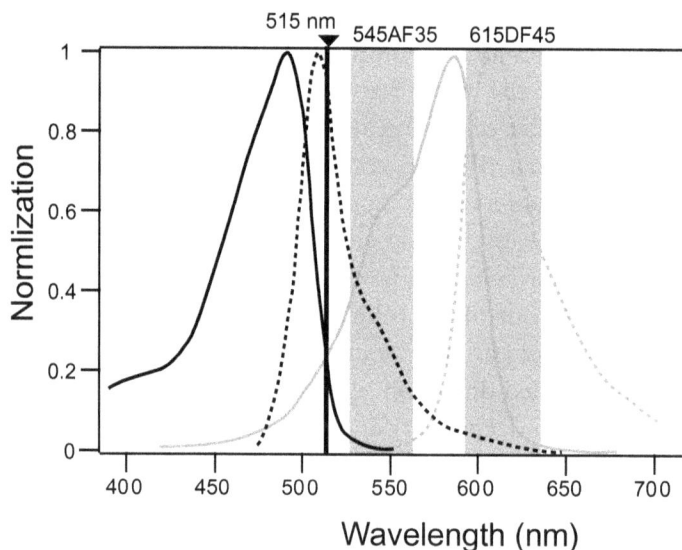

**Fig. 3** Excitation and emission spectra of the fluorophore pair of EGFP and mRFP, shown together with emission filters. It is feasible to excite both EGFP and mRFP at 515 nm and separately collect fluorescence signals at 545 nm for EGFP and 615 nm for mRFP, using emission filters of 545AF35 and 615DF45, respectively. The black lines are for EGFP and the grey lines are for mRFP. The solid lines and dotted lines represent excitation and emission spectra, respectively.

dyes and FPs with exceptionally large Stokes shift have been reported recently,[10,64] which greatly improved the applicability of SW-FCCS both *in vitro* and *in vivo*. SW-FCCS in general has simplified experimental procedures, reasonable signal-to-noise ratio, and can be performed with readily available instruments. One example of the SW-FCCS application in living zebrafish embryos will be discussed in the following section.

In biology, investigation of living cells nowadays relies heavily on the use of FPs as fluorescent tags. Aside from their low molecular brightness and low photo-stability, FPs also suffer from spectral cross talk due to their long-tailed emission. In FCCS measurements, spectral cross-talk will produce false-positive cross-correlation signals and makes quantification difficult. Pulsed interleaved excitation (PIE), which utilizes two excitation lasers that are pulsed alternatively, was therefore implemented in FCCS for this issue.[65] Two pulsed laser beams are superimposed and synchronized to excite two fluorophores alternatively in PIE. A single detector is used to count the emitted photon with a time tag and the pulse repetition rate is carefully controlled to allow for fluorescence decay after excitation. Off-line cross-correlation analysis of photons collected after each pulse produces cross-correlation functions that are free of spectral cross talk.

In FCS, in order to measure the absolute diffusion coefficient, one has to know precisely the shape and size of the observation volume (as discussed above, the parameter directly obtained from FCS is diffusion time $\tau_d$, the diffusion coefficient $D$ is calculated by $\tau_d = \omega_0^2/4D$). Practically, the diffusion coefficient is indirectly determined by a calibration measurement using dyes with known diffusion coefficients. However, the shape and size of the observation volume depends on the experimental conditions such as cover glass thickness variation and laser beam characteristics. When working in a multicellular living organism, the refractive index mismatch between water and tissue will also lead to a distortion of the observation volume and the distortion is dependent upon the penetration depth. Thus a quantitative determination using a calibration approach in this case is problematic. This problem can be overcome by two-focus FCCS, which employs two spatial separated observation volumes with a well-defined and known distance.[66,67] This introduces an external ruler into the measurement (the known distance), which is absent in conventional FCS. Fluorescence signals from the two observation volumes are auto- and cross-correlated, and the characteristic time which requires the fluorophore to diffuse from one observation volume to the other can be determined. The distance between the two observation volumes is not changed by refractive index mismatch, or cover glass thickness variation, or laser beam characteristic, thus an absolute diffusion coefficient can be directly obtained from the measurements, without additional calibration measurements. Two-focus FCCS therefore can effectively eliminate the problem of observation volumes changes encountered in biological samples.

Another challenge of FCS/FCCS applications in living cells and living embryos is the correction for cell, especially cell membrane, movements during sample measurements. In living zebrafish embryos, cells divide every 15 minutes during the first couple of hours of embryogenesis and later on the cell cycle slows down and often is longer than one–two hours, and some cell movements are relatively fast, but some other are quite slow. One FCS/FCCS measurement in a living cell takes 30 seconds or longer, during which time a cell or cell membrane movement will complicate the statistical analysis, especially for FCCS which determines binding based on the concurrent movements of two fluorescent labels. Cell or cell membrane movement will result in a false positive cross-correlation signal, as different fluorescent labels even when not interacting appear to be moving together. A solution to this problem is to chose carefully cell population to study and/or scan the observation volume.[68] In scanning FCS, the observation volume is repeatedly scanned across the sample in a controlled way while fluorescence signals are collected. The correlation analysis can be performed from all points along the trajectory or from selected ones only. In principle, any dynamics slower than the repetition rate of the scan can be efficiently determined. In scanning FCS, membrane movements can be corrected by scanning the laser beam across the membrane perpendicular to the membrane surface. Calculating the position of the maximum fluorescence intensities along each single line, corresponding to the actual membrane position, and correlating only signals from these positions can compensate movements of the membrane off-line. In addition, scanning FCS distributes the laser power over a larger area, which reduces fluorophore photo-bleaching in the observation volume. This allows measurements of slowly diffusing particles.

## Application of FCS in Zebrafish Embryos

The zebrafish, *Danio rerio*, is increasingly popular as a research animal model in various research fields. In particular, researchers are more aware that zebrafish can be the vertebrate model of choice for cell biological-based research.[69] This is facilitated by the optical transparency of the zebrafish embryos and early larvae, which allows direct visualization of cells and organs deep within tissue. The transparent zebrafish embryo body means less light blockage by the tissue in the visible spectrum to human eyes, which is 380 to 750 nm in wavelength. This wavelength range covers the excitation and emission spectra of most commonly used fluorescent molecules, aiding the application of fluorescence microscopy and spectroscopy. Over the past three years, several publications have demonstrated the viability of direct investigation of molecular dynamics and interactions in living zebrafish embryos using FCS and its variants.[26,70–74] Several examples will be presented in this section, including measurements of blood flow velocities, diffusion coefficients of cytosolic and membrane-bound proteins, and dissociation constants of interacting proteins.

As conventional FCS utilizes confocal illumination, the sample preparation of zebrafish embryos for FCS measurements is similar to that of confocal microscopy. Figure 4 shows one example of sample mounting in an inverted microscope setup. Fluorescent tags can be introduced by microinjecting mRNAs or DNAs into one blastomere at 1–32 cells stage, or by using transgenic embryos. The mRNA in general has the advantage of earlier formation of fluorescent proteins, their more uniform distribution in different cell populations and more precise control of the expression level. But mRNA is short lived and

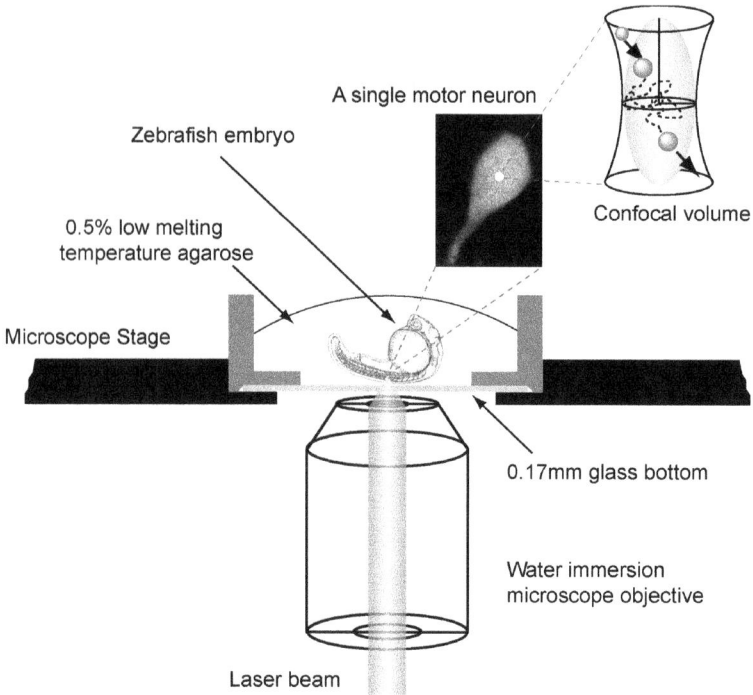

**Fig. 4** Preparation of zebrafish embryos for FCS measurements. Selected embryos are anesthetized and mounted into 0.5% low melting temperature agarose in a glass bottom Petri dish. The embryo is positioned close to the cover glass and the orientation of the embryo can be adjusted to fit specific applications. Here, the embryo was mounted for dorsal views. The specimen was then placed on the microscope stage for examination. Use of a water immersion microscope objective with high magnification and numerical aperture allows observation of single cells within the embryo body, for example a single motor neuron cell. With the help of the confocal image, the observation volume can be placed within the cell for subsequent FCS measurement.

cannot sustain protein expression to later stages. On the other hand, DNA containing EGFP genes need more time for maturation but the expression lasts longer. Depending on the embryonic stages of interest, mRNA and DNA can be chosen accordingly. One challenge for FCS application in the zebrafish embryo is the efficient suppression of pigmentation during development. Melanin, the pigment found in the epidermal layer of skin, has a large scattering coefficient in the UV region, which protects the skin from damaging UV radiation of sun. It also has a significant absorption coefficient in the NIR region. Therefore, it is preferable to work with embryos of fish strains where pigmentation is suppressed genetically (e.g. *albino*) or chemically (PTU, 0.003% 1-phenyl-2-thiourea in 10% Hank's saline). Selected embryos are anesthetized first to prevent body movement, and subsequently mounted in 0.5%–1% agarose. The zebrafish embryo can survive in agarose more than a day so FCS measurements can be repeatedly performed during development of the same embryo. For FCS application in living cells and tissues, use of a water immersion microscope objective with high magnification and numerical aperture is preferred. One drawback of this kind of objective is the short working distance, usually less than 300 μm. So the embryo should be mounted close to the cover glass. After embryo placement on the microscope stage, confocal imaging is performed first to guide positioning of the observation volumes to a specific subcellular compartment. Modern FCM usually use a single pinhole for both imaging and FCS detection.[38] This guarantees the accurate position of the observation volume after confocal image acquisition.

FCS can be used to measure blood flow velocities and directions in living zebrafish embryos.[70-72] Blood flow measurements play an important role in studies of vascular development. Measurements of blood flow velocities in living animals provide insights in studying the shear stress induced upon vessel walls and understanding the development of vascular endothelial cells. Previous methods, such as imaging analysis by fast confocal laser scanning microscopy,[75] provide data characterized by good temporal but low spatial resolution. However, a high level of spatial resolution is a key factor to understand the local changes of blood flow in the vessel. In this regard, FCS is an attractive alternative to measure flow. Firstly, the spatial resolution of FCS relies on the size of the observation volume, which is about 0.5 μm in diameter. This value is considerably smaller than the size of the zebrafish embryo blood vessels. FCS thereby allows determination of the spatial flow profile across the vessel and shear stress along the vessel wall can be studied in detail.[74] Secondly, FCS works at extremely low concentrations of small dye molecules on the range of 1 nM, eliminating the problem of vessel obstruction by large probes required by other methods. In fact, it has been shown that the autofluorescence present in the serum generates sufficient fluorescence signal for FCS analysis.[72] No additional dye injection into the blood circulation is required, which makes FCS a true noninvasive approach. Since this can be accomplished based on the autofluorescence of serum, FCS can also be used to measure blood flow in small vessels during embryogenesis, when red blood cells are not yet present, or in mutants, where formation of blood cells has been affected. Figure 5 shows one example of blood flow velocity measurements in the dorsal aorta and cardinal vein. A three days post-fertilization (dpf) wild type zebrafish embryo is mounted in lateral view, and FCS measurements were performed using 100 µW of a 543 nm laser beam. Prior to the measurement, the embryo is anesthetized by 500 µM of tricaine. However, tricaine at high dose is known to suppress cardiac function and the effect can vary from embryo to embryo. Thus velocity comparison is only conducted within one fish. As the flow velocities are not equal across the vessel, the observation volumes are positioned at the center of the vessel to account for the maximum velocity. A typical ACF curve taken in the cardinal vein is shown in Fig. 5B. A fitting function for laminar flow has been previously proposed for FCS measurements.[76] However, due to the dynamic change of flow velocities in living animals, a full resolution of the velocity profile is not possible with the current fitting model. In general, the blood flow velocity can be divided into two alternating fast and slow portions, which are induced by the heart contraction (systole) and heart relaxation (diastole) in the cardiac cycle. Therefore a simplified two-flow model (Fig. 5B) has been used to account for the average systolic and diastolic flow velocities.[72] This can be viewed as the mimic of using systolic and diastolic pressure to describe blood pressures on vessel walls. Fig. 5C shows the normalized ACF curves obtained from dorsal aorta and cardinal vein, and the results suggest that the velocities of both systolic and diastolic flow in the dorsal aorta are faster than those in the cardinal vein. This is expected due to the different levels of cardiac-mediated blood pressure in these vessels and their vessel width (the dorsal aorta is ~15 µm in width and the cardinal vein

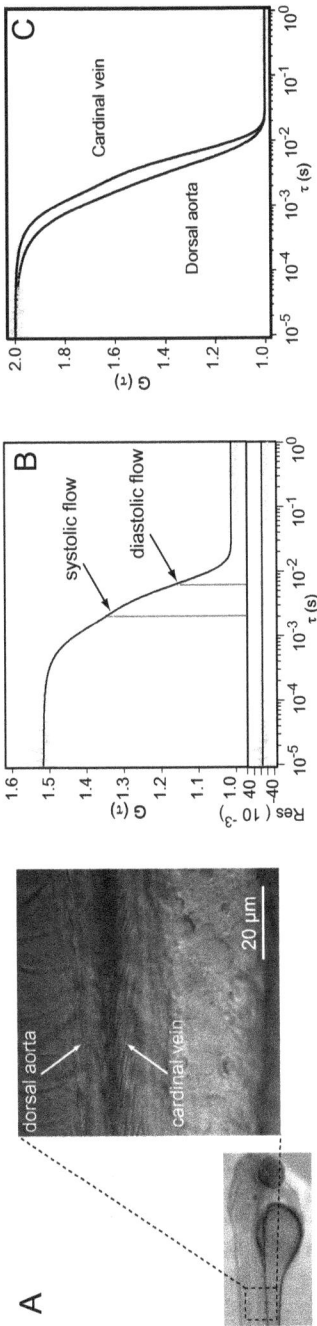

**Fig. 5** Example of blood flow measurements in living zebrafish embryos using FCS. (**A**) Transmission channel cofocal images showing the dorsal aorta and cardinal vein of a three dpf wild type zebrafish embryos. (**B**) A typical FCS measurement of blood flow, showing both experimental curve (grey) and its fit (black). The experimental curve is fitted with a two-flow speed model, representing the average systolic and diastolic flows. At the bottom are the residuals of the fit, a measure for the quality of the fit. (**C**) Normalized FCS blood flow measurements obtained from dorsal aorta and cardinal vein, shown together with experimental curves (grey) and fitting curves (black).

is ~19 μm). It should be noted that FCS blood flow measurements should avoid regions close to the heart as the body movement induced by heartbeat introduces an additional periodic time trace into the ACF making analysis difficult.

The accurate determination of the diffusive behavior of proteins provides evidence for and new insights into protein functions. FCS is particularly suitable for this task as it noninvasively characterizes molecular dynamics from milliseconds to seconds, which covers the dynamic range of most biomolecules in living cells. We show here one example of diffusion coefficient measurements of EGFP in cytoplasm and nucleoplasm in single motor neurons within a living zebrafish embryo (unpublished data). The physical structures of cell cytoplasm and nucleoplasm have been an interesting topic for decades, and it is clear that cytoplasm and nucleoplasm have distinct properties. The cytoplasm comprises dissolved solutes and macromolecules in a complex array of microtubules, actin, and intermediate filaments; and it contains various subcellular compartments. The nucleoplasm, in comparison, is mainly filled with chromatin whose structure is dynamically changing; and it contains no subcellular compartment with membrane structure. Numerous publications have studied the translational diffusion of fluorescent probes or functional proteins in cytoplasm and nucleoplasm, aiming to study the molecular crowding of the interior of living cells, size-dependent DNA or protein mobility, protein distributions and functions, and chromatin densities. In particular, two recent publications have suggested that EGFP diffuses faster in the cytoplasm than in the nucleoplasm.[77,78] Those experiments were performed in various cell cultures. However, cell cultures with unfavorable cell status may introduce potential artifacts and compromise the interpretation of the results. For example, the cell nucleus can undergo dramatic structural rearrangement during the cell division cycle, e.g. at synthesis stage (S phase) DNA replication could increase the chromatin amount and subsequently the nucleus will become more crowded. Thus translational diffusion measurements in cell cultures at an unfavorable phase will lead to biased data. In comparison, a living organism provides another level of control not possible in artificial medium buffers and Petri dishes. The motor neuron cells for instance usually stay indefinitely at $G_0$ phase and will not divide anymore after maturation, which makes it a good model to study translational diffusion of biomolecules in the nucleus. Figure 6 shows two sets of diffusion time measurements in the cytoplasm and nucleo-

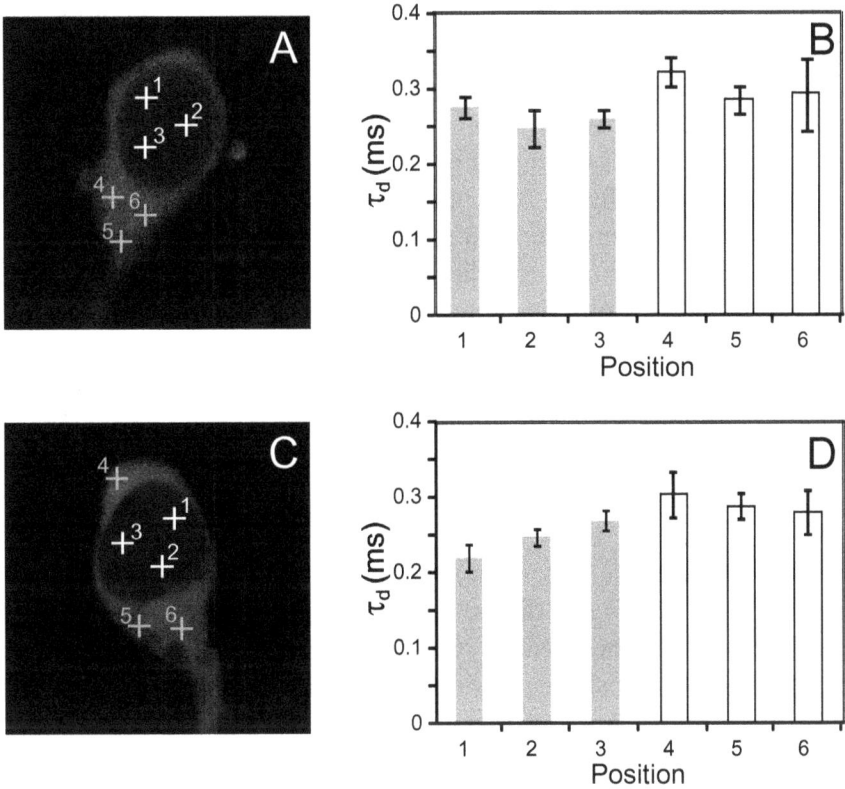

**Fig. 6** Diffusion time measurements in cytoplasm and nucleoplasm within one cell. Motor neuron cells of a three dpf *Islet-1*-EGFP transgenic embryo are chosen for this purpose. Due to the small size, EGFP molecules are evenly distributed in cytoplasm and nucleoplasm. The nucleus is outlined by performing point photobleaching in the nucleus. (**A**) and (**C**) are the confocal images captured right after the point photobleaching. As EGFP molecules cannot diffuse through the nuclear pore fast enough, the fluorescence intensity is lower in the nucleus than in the cytoplasm. Three points were selected in both nucleus and cytoplasm for FCS measurements as indicated by the crosses, and the obtained diffusion times are plotted as shown in (**B**) and (**D**), respectively.

plasm within single motor neurons. The *Islet-1*-EGFP transgenic line is used for this purpose. In this transgenic line, the neural-specific *Islet-1* promoter/enhancer drives EGFP expression in primary neurons, including a subset of sensory neurons and motor neurons.[79] Three points were selected in both cytoplasm and nucleoplasm for FCS measurement as marked in Figs. 6A and 6C, and four measurements were performed in each point. The obtained diffusion time is plotted in Figs. 6B and 6D. Interestingly, the results show that EGFP diffuses slightly slower in the cytoplasm than in the nucleoplasm, contrary to previous reports conducted in cell cultures. FCS measurements were repeated in several more motor neurons in different zebrafish embryos and the average diffusion time obtained in cytoplasm and nucleoplasm were $0.29 \pm 0.04$ ms and $0.23 \pm 0.03$ ms, respectively (18 measurements each), confirming the slightly slower diffusion of EGFP in the cytoplasm. The results also suggest that the EGFP molecules diffuse relatively fast in both cytoplasm and nucleoplasm, considering the macromolecular crowding environment within cells. Measurements of EGFP in buffer solution at the same experimental condition yielded a diffusion time of $0.19 \pm 0.02$ ms, which is only ~40% faster than in the cytoplasm. This could be a consequence of the relative small size of EGFP (27 kDa), as organelles, cytoskeleton, and even dissolved macromolecules are less obstructive for small-sized probe and the EGFP molecule can still diffuse fast within aqueous voids.

FCS can also be used to study membrane-bound receptor proteins. Here we show one example of diffusion coefficient measurements of a seven transmembrane domain receptor protein Cxcr4b (CXC chemokine receptor 4b) in living zebrafish embryos.[74] Cxcr4b is a chemokine receptor that belongs to the G-Protein Coupled Receptor (GPCR) gene family.[80] These receptors play an important role in defining directionality of cell migration of germ cells, lateral line, and somites during embryogenesis.[81] In this experiment, EGFP was attached to the C-terminus of Cxcr4b (Fig. 7A) to avoid obstruction of ligand binding to Cxcr4b at the N-terminus. The constructed Cxcr4b-EGFP plasmid was microinjected into a single blastomere at 16-cell stage. As shown by confocal imaging (Fig. 7B), the EGFP fluorescence signals were found along the membrane of

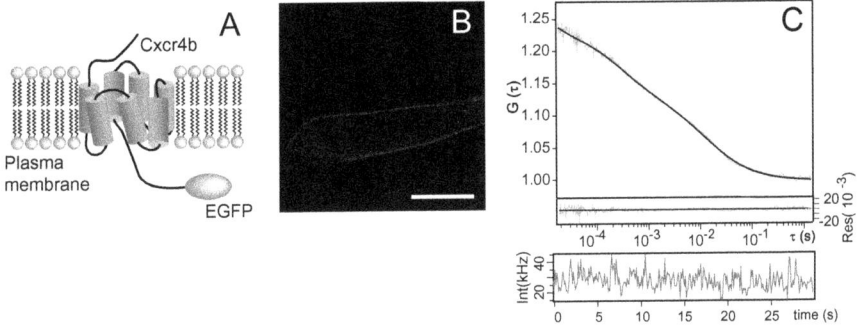

**Fig. 7** Diffusion time measurements of the EGFP labeled transmembrane receptor protein Cxcr4b. **(A)** Schematic drawing of the seven-pass transmembrane receptor, tagged with EGFP at the N terminus. **(B)** Confocal image of one Cxcr4b-EGFP expressing muscle fiber. Scale bar = 20 $\mu$m. **(C)** One example of FCS measurement of Cxcr4b-EGFP on membrane, showing experimental curve (grey) and its fit (black). Beneath are residuals of the fit and the fluorescence intensity trace.

muscle fiber cells and the expression was uniformly distributed. Muscle fiber cells were chosen for this measurement as they are abundant within the embryo and are easy to identify. Low concentration of Cxcr4b-EGFP DNA was injected as it has been shown that over-expression of Cxcr4b leads to developmental abnormalities,[82] and low fluorescent probe concentration is more suitable for FCS application. In FCS application, the orientation and topography of the cell membrane can greatly influence FCS measurements. Therefore correct positioning of the observation volume on the membrane is critical for accurate data acquisition.[83] The most commonly used method is to position the volume on the top or bottom membranes (parallel to the *x-y* plane) and adjust the height of the volume to achieve highest fluorescence intensity. In this experiment, the FCS observation volume was placed on the bottom membrane of the muscle fiber cell (more superficial to the embryo surface) and the *z*-position was adjusted in steps of 0.1 μm until the maximum fluorescence intensity was observed. A typical ACF curve is shown in Fig. 7C. The fluorescence intensity trace showed no intensity bursts, suggesting no higher order aggregation of Cxcr4b on the membrane. The fitting yielded a diffusion time of 22.1 ± 7.1 ms for Cxcr4b-EGFP on the membrane. After a calibration measurement of a chemical dye with known diffusion coefficient, the diffusion coefficient of Cxcr4b-EGFP is estimated to be $0.60 \pm 0.19 \ \mu m^2 s^{-1}$, in agreement with previously published results obtained in cell cultures.[84] It should be noted that Cxcr4b was found to form homodimers and heterodimers with Cxcr7 in response to ligand binding.[85,86] Thus the diffusion coefficient obtained here depicts the diffusion behavior of the homo- and hetero-dimer complex.

The last example we present is the determination of the dissociation constant of an interacting protein pair in living zebrafish embryos with SW-FCCS. The quantification of biological interactions plays important role in understanding cell behavior and designing drug molecules. Recent advances in molecular and cell biology show that protein-protein interactions, at the molecular level, are the essence of most biological processes. Conventional methods for binding quantification usually require a simplified and controllable environment. Consequently the quantification of biomolecular interactions is performed in buffer solutions and at best in Petri dish-based cell cultures. However, in the last years, evidence emerged which suggests that those *in vitro* findings do not necessarily correlate with values in physiological tissues. Furthermore, biomolecular interactions also strongly depend on, for example, the concentration of the reactant and the 3D structure of the multiple cell layers. It is therefore increasingly important to quantify biological interactions in an intact organism to obtain physiological relevant parameters. SW-FCCS, as discussed in the previous section, is capable of quantifying protein-protein interactions in living cells and the application can be extended to multicellular organisms due to the use of single light source. We have reported the determination of the dissociation constant between Cdc42 (cell division cycle 42) and IQGAP1 (IQ motif containing GTPase activating protein 1) with SW-FCCS in living zebrafish embryos.[26] Cdc42 is a small GTPase that belongs to the Rho/Rac subfamily. Its interaction with IQGAP1, a multi-domain protein with several protein-interacting motifs, has been previously confirmed both *in vitro* and *in vivo*.[87] The interaction of IQGAP1 and Cdc42 plays an important role in modeling microtubules and the cytoskeleton during cell polarization and migration.[88] Two Cdc42 mutants were used in this work: the constitutively active $Cdc42^{G12V}$, which is in a predominantly GTP-bound form and the dominant-negative GDP-bound $Cdc42^{T17N}$. The EGFP and mRFP genes were attached to IQGAP1 and Cdc42 DNA respectively and the constructed plasmids were co-injected into zebrafish embryos at the 16-cell stage. Figure 8A shows the confocal image of one muscle fiber cell expressing both EGFP-IQGAP1 and the active mutant $mRFP-Cdc42^{G12V}$. SW-FCCS measurements were performed in the cytoplasm and one example is shown in Fig. 8B. It is obvious from the SW-FCCS result that the two proteins form complexes, as shown by the elevated CCF amplitude. The concentrations of both bound ($C_{gr}$) and free ($C_g$ for EGFP-IQGAP1 and $C_r$ for $mRFP-Cdc42^{G12V}$) proteins can be calculated from the ACF and CCF functions.[26] Therefore a scattering plot of the product $C_g \times C_r$ against $C_{gr}$ can be used to calculate the dissociation constant $K_D$ ($K_D = C_g \times C_r/C_{gr}$) as shown in Fig. 8C. The $K_D$ value determined using this method is $105 \pm 11$ nM, also suggesting a strong interaction. In comparison, Figs. 8D–8F show the result of the protein pair of EGFP-IQGAP1 and the dominant-negative mutant mRFP-$Cdc42^{T17N}$. The CCF curve suggests no or weak binding and the $K_D$ value calculated is ~1500 nM. The same experiments were also performed in CHO cells.[26] Interestingly, the $K_D$ value for the protein pair of EGFP-IQGAP1 and the active mutant $mRFP-Cdc42^{G12V}$ is ~1000 nM, which is much higher than that in the zebrafish muscle fibers, while data on interaction of the dominant negative mutant protein are similar. This suggests that even the constitutively active $Cdc42^{G12V}$ interacts weakly with IQGAP1 in CHO cells. One possible explanation is the $Ca^{2+}$ concentration-dependent interactions of IQGAP1 and Cdc42. Several reports have shown that increased intracellular $Ca^{2+}$ concentration will lead to binding of calmodulin to IQGAP1, which in turn promotes the dissociation of IQGAP1 from Cdc42.[89] The CHO cells were cultured in a commonly used F-12K Kaighn's modification medium which includes 1 mM calcium chloride and other calcium containing chemicals. Therefore the intracellular $Ca^{2+}$ concentration in CHO cells could be different from the level found in zebrafish tissues, which caused the weak interaction between IQGAP1 and $Cdc42^{G12V}$.

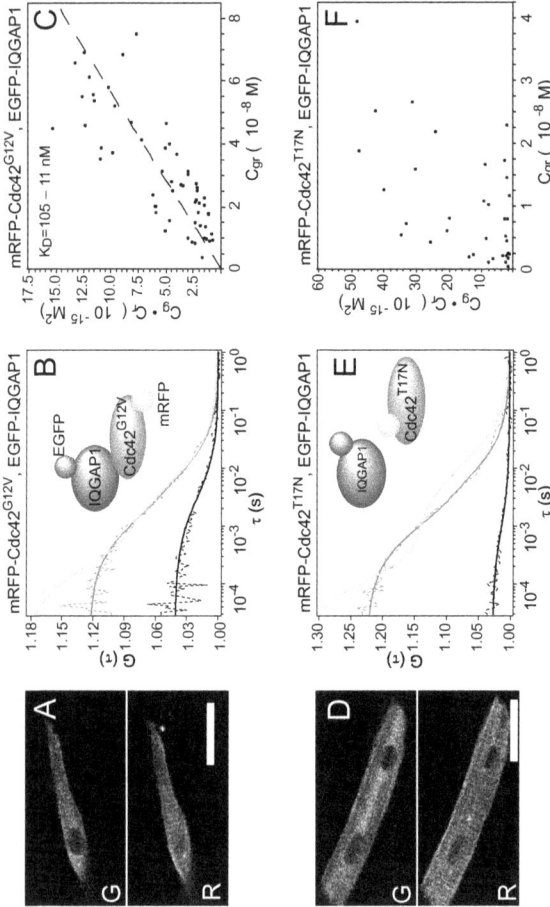

**Fig. 8.** SW-FCCS measurements of the interaction of Cdc42 and IQGAP1. (**A**) Confocal images of one muscle fiber that expresses mRFP-Cdc42$^{G12V}$ (R for red channel), an active mutant of Cdc42, and EGFP-IQGAP1 (G for green channel). Scale bar = 20 $\mu$m. (**B**) SW-FCCS result of the interacting protein pair, showing both experimental curves (dotted line) and their fits (solid line). The light grey, dark grey and black represent the red ACF, green ACF and complex CCF curves, respectively. The amplitude of the CCF curve is high indicating binding. The insert is the schematic drawing of the sample. (**C**) Determination of dissociation constant $K_D$ for the interacting protein pair using a scattering plot. The concentration of the red particle ($C_r$), green particle ($C_g$) and the complex ($C_{gr}$) can be obtained from SW-FCCS measurement, and $C_g \cdot C_r$ is plotted against $C_{gr}$ to obtain $K_D$ ($K_D = C_g \cdot C_r/C_{gr}$). The obtained $K_D$ for the interaction of Cdc42$^{G12V}$ and IQGAP1 is 105 ±11nM, indicating strong binding. (**D**), (**E**) and (**F**) are the corresponding figures for the protein pair of Cdc42$^{T17N}$, a dominant negative mutant of Cdc42, and IQGAP1. The amplitude of the CCF curve is low indicating no or weak binding. The scatter plot suggests no correlation but when fitted results in a $K_D$ value ~1500 nM, suggesting very weak binding.

# Conclusion

Over the past few years, fluorescence-based biophysical techniques, e.g. fluorescence recovery after photobleaching (FRAP), FRET, fluorescence life time imaging microscopy (FLIM) and FCS have been established as routine tools to probe biomolecular dynamics and interactions in Petri dish-based cell cultures. However, how closely *in vitro* findings reflect *in vivo* biological process is unclear. Cell cultures are engineered as isolated individual cells that can be artificially cultivated. Nevertheless, the flat glass substrate and the artificial medium buffer are significantly different from a real physiological environment. The absence of extracellular matrix and various cell-cell communication functions also makes the information collected *in vitro* not predictive for drug development. For instance, it has been shown that even genetically normal primary cells placed in cell culture quickly lose their differentiated gene expression pattern and phenotype.[90] In addition, many biological question, especially those of developmental processes, cannot be addressed in 2D cell cultures but necessarily require the work in living animals. It is therefore desirable to study biomolecules directly in an *in vivo* system.

Untill now, FCS application in living animals is still limited. The multiple cell layers in thick tissue hinder many fluorescence-based investigations. Nevertheless, using different approaches several examples of application of FCS in living animals can be found in literature. Nagao *et al.* reported diffusion coefficient measurements of GFP labeled granules in medaka primordial germ cells using FCS and FRAP.[91] To avoid the deep tissue penetration, the medaka embryos were dissected and the sectioned tissue was placed on a cover glass to reveal cells of interest. The same approach was adopted when *Drosophila melanogaster* was used as a model for FCS applications.[74] Another obvious way to avoid the problem of deep tissue penetration is to study the surface cells of an animal or use smaller animal embryos with a limited number of cell layers. Petrasek *et al.* applied scanning FCS to study the localization and redistribution of GFP labeled NMY-2 and PAR-2 proteins during asymmetric first division of *Caenorhabditis elegans* embryos representing invertebrates.[92] We presented here several examples illustrating a possibility to use the transparent vertebrate embryos to alleviate this problem. Molecular dynamics and interactions can be directly measured in living zebrafish embryos with FCS and its variants. With the new developments in fluorescence spectroscopy, and in particular in FCS, single molecule measurements could be performed using small model vertebrate species, which provide the developmental biologist with a new set of noninvasive tools to address questions related to interaction of molecules *in vivo*.

# References

1. Hirschfeld T. Optical microscopic observation of single small molecules. *Appl Opt* 15:2965–2966, 1976.
2. Lakowicz JR. Principles of Fluorescence Spectroscopy. Springer, 2006.
3. Liu P, Ahmed S and Wohland T. The F-techniques: advances in receptor protein studies. *Trends Endocrinol Metab* 19:181–190, 2008.
4. Magde D, Elson E and Webb WW. Thermodynamic fluctuations in a reacting system-measurement by fluorescence correlation spectroscopy. *Phys Rev lett* 29:705–708, 1972.
5. Krichevsky O and Bonnet G. Fluorescence correlation spectroscopy: the technique and its applications. *Rep Prog Phys* 65:251–297, 2002.
6. Chalfie M, Tu Y, Euskirchen G, Ward WW and Prasher DC. Green fluorescent protein as a marker for gene expression. *Science* 263:802–805, 1994.
7. Ormo M, Cubitt AB, Kallio K, Gross LA, Tsien RY, and Remington SJ. Crystal structure of the *Aequorea victoria* green fluorescent protein. *Science* 273:1392–1395, 1996.
8. Shimomura O, Johnson FH and Saiga Y. Extraction, purification and properties of aequorin, a bioluminescent protein from the luminous hydromedusan, *Aequorea*. *J Cell Comp Physiol* 59:223–239, 1962.
9. Shaner NC, Campbell RE, Steinbach PA, Giepmans, Palmer AE, and Tsien RY. Improved monomeric red, orange and yellow fluorescent proteins derived from *Discosoma* sp. red fluorescent protein. *Nat Biotechnol* 22:1567–1572, 2004.
10. Shcherbo D, Merzlyak EM, Chepurnykh TV, Fradkov AF, Ermakova GV, Solovieva EA, Lukyanov KA, Bogdanova EA, Zaraisky AG, Lukyanov S, and Chudakov EM. Bright far-red fluorescent protein for whole-body imaging. *Nat Methods* 4:741–746, 2007.

11. Widengren J and Rigler R. Fluorescence correlation spectroscopy as a tool to investigate chemical reactions in solutions and on cell surfaces. *Cell Mol Biol* 44:857–879, 1998.

12. Rauer B, Neumann E, Widengren J, and Rigler R. Fluorescence correlation spectrometry of the interaction kinetics of tetramethylrhodamin alpha-bungarotoxin with Torpedo californica acetylcholine receptor. *Biophys Chem* 58:3–12, 1996.

13. Van Craenenbroeck E and Engelborghs Y. Quantitative characterization of the binding of fluorescently labeled colchicine to tubulin *in vitro* using fluorescence correlation spectroscopy. *Biochemistry* 38:5082–5088, 1999.

14. Wohland T, Friedrich K, Hovius R and Vogel H. Study of ligand-receptor interactions by fluorescence correlation spectroscopy with different fluorophores: evidence that the homopentameric 5-hydroxytryptamine type 3As receptor binds only one ligand. *Biochemistry* 38:8671–8681, 1999.

15. Starchev K, Zhang J and Buffle J. Applications of fluorescence correlation spectroscopy — particle size effect. *J Colloid Interface Science* 203:189–196, 1998.

16. Pramanik A and Rigler R. Ligand-receptor interactions in the membrane of cultured cells monitored by fluorescence correlation spectroscopy. *Biol Chem* 382:371–378, 2001.

17. Kahya N, Scherfeld D, Bacia K and Schwille P. Lipid domain formation and dynamics in giant unilamellar vesicles explored by fluorescence correlation spectroscopy. *J Struct Biol* 147:77–89, 2004.

18. Bacia K, Scherfeld D, Kahya N and Schwille P. Fluorescence correlation spectroscopy relates rafts in model and native membranes. *Biophy J* 87:1034–1043, 2004.

19. Eggeling C, Ringemann C, Medda R, Schwarzmann G, Sandhoff K, Polyakova S, V. Belov N, Hein B, von Middendorff C, Schonle A and Hell SW. Direct observation of the nanoscale dynamics of membrane lipids in a living cell. *Nature* 457:1159–1162, 2009.

20. Briddon SJ and Hill SJ. Pharmacology under the microscope: the use of fluorescence correlation spectroscopy to determine the properties of ligand-receptor complexes. *Trends Pharmacol Sciences* 28:637–645, 2007.

21. Bacia K and Schwille P. Practical guidelines for dual-color fluorescence cross-correlation spectroscopy. *Nat Protoc* 2:2842–2856, 2007.

22. Hwang LC and Wohland T. Recent advances in fluorescence cross-correlation spectroscopy. *Cell Biochem Biophys* 49:1–13, 2007.

23. Thompson NL, Lieto AM and Allen NW. Recent advances in fluorescence correlation spectroscopy. *Curr Opin Struct Biol* 12:634–641, 2002.

24. Kim SA, Heinze KG and Schwille P. Fluorescence correlation spectroscopy in living cells. *Nat Methods* 4:963–973, 2007.

25. Liu P, Ahmed S, and Wohland T. The F-techniques: advances in receptor protein studies. *Trends Endocrinol Metabo* 19:181–190, 2008.

26. Shi X, Foo YH, Sudhaharan T, Chong SW, Korzh V, Ahmed S and Wohland T. Determination of dissociation constants in living zebrafish embryos with single wavelength fluorescence cross-correlation spectroscopy. *Biophys J* 97:678–686, 2009.

27. Magatti D and Ferri F. 25 ns software correlator for photon and fluorescence correlation spectroscopy. *Rev Sci Instrum* 74:1135, 2003.

28. Schatzel K, Drewel M and Stimac S. Photon correlation measurements at large lag times: improving statistical accuracy. *J Mod Opt* 35:711–718, 1988.

29. Wahl M, Gregor I, Patting M and Enderlein J. Fast calculation of fluorescence correlation data with asynchronous time-correlated single-photon counting. *Opt Express* 11:3583–3591, 2003.

30. Ricka J and Binkert T. 1989. Direct measurement of a distinct correlation function by fluorescence cross correlation. *Phys Rev* 39:2646–2652, 1989.

31. Rigler R, Mets Üidengren J and Kask P. Fluorescence correlation spectroscopy with high count rate and low background: analysis of translational diffusion. *Eur Biophys J* 22:169–175, 1993.

32. Koppel DE. Statistical accuracy in fluorescence correlation spectroscopy. *Phys Rev A* 10:1938–1945, 1974.

33. Aragon SR and Pecora R. Fluorescence correlation spectroscopy as a probe of molecular dynamics. *J Chem Phys* 64:1791, 1976.

34. Shi X and Wohland T. Fluorescence correlation spectroscopy. In: *Nanoscopy and Multidimensional Optical Fluorescence Microscopy.* Diaspro A (ed.) CRC Press, 2010.

35. Kastrup L, Blom H, Eggeling C and Hell SW. Fluorescence fluctuation spectroscopy in subdiffraction focal volumes. *Phys Rev Lett* 94:178104, 2005.

36. Estrada LC, Aramendia PF and Martinez OE. 10000 times volume reduction for fluorescence correlation spectroscopy using nanoantennas. *Opt Express* 16:20597–20602, 2008.

37. Starr TE and Thompson NL. Total internal reflection with fluorescence correlation spectroscopy: combined surface reaction and solution diffusion. *Biophys J* 80:1575–1584, 2001.
38. Pan X, Foo W, Lim W, Fok MH, Liu P, Yu H, Maruyama I and Wohland T. Multifunctional fluorescence correlation microscope for intracellular and microfluidic measurements. *Rev Sci Instrum* 78:053711, 2007.
39. Wruss J, Runzler D, Steiger C, Chiba P, Kohler G and Blaas D. Attachment of VLDL receptors to an icosahedral virus along the 5-fold symmetry axis: multiple binding modes evidenced by fluorescence correlation spectroscopy. *Biochemistry* 46:6331–6339, 2007.
40. Meseth U, Wohland T, Rigler R and Vogel H. Resolution of fluorescence correlation measurements. *Biophys J* 76:1619–1631, 1999.
41. Eigen M and Rigler R. Sorting single molecules: application to diagnostics and evolutionary biotechnology. *Proc Nat Acad Sci USA* 91:5740–5747, 1994.
42. Schwille P, Meyer-Almes JF and Rigler R. Dual-color fluorescence cross-correlation spectroscopy for multicomponent diffusional analysis in solution. *Biophys* Journal 72:1878–1886, 1997.
43. Camacho A, Korn K, Damond M, Cajot JF, Litborn E, Liao B, Thyberg P, Winter H, Honegger A, Gardellin P and Rigler R. Direct quantification of mRNA expression levels using single molecule detection. *J Biotechnol* 107:107–114, 2004.
44. Foldes-Papp Z and Rigler R. Quantitative two-color fluorescence cross-correlation spectroscopy in the analysis of polymerase chain reaction. *Biol Chem* 382:473–478, 2001.
45. Kettling U, Koltermann A, Schwille P and Eigen M. Real-time enzyme kinetics monitored by dual-color fluorescence cross-correlation spectroscopy. *Proc Nat Acad Sci USA* 95:1416–1420, 1998.
46. Korn K, Gardellin P, Liao B, Amacker M, Bergstrom A, Bjorkman H, Camacho A, Dorhofer S, Dorre K, Enstrom J, Ericson T, Favez T, Gosch M, Honegger A, Jaccoud S, Lapczyna M, Litborn E, Thyberg P, Winter H and Rigler R. Gene expression analysis using single molecule detection. *Nucleic Acids Res* 31:e89, 2003.
47. Bacia K, Majoul IV and Schwille P. Probing the endocytic pathway in live cells using dual-color fluorescence cross-correlation analysis. *Biophys J* 83:1184–1193, 2002.

48. Baudendistel N, Muller G, Waldeck W, Angel P and Langowski J. Two-hybrid fluorescence cross-correlation spectroscopy detects protein-protein interactions *in vivo. Chemphyschem* 6:984–990, 2005.
49. Muto H, Nagao I, Demura T, Fukuda H, Kinjo M and Yamamoto KT. Fluorescence cross-correlation analyses of the molecular interaction between an Aux/IAA protein, MSG2/IAA19, and protein-protein interaction domains of auxin response factors of arabidopsis expressed in HeLa cells. *Plant Cell physiol* 47:1095–1101, 2006.
50. Saito K, Wada I, Tamura M and Kinjo M. Direct detection of caspase-3 activation in single live cells by cross-correlation analysis. *Biochem Biophys Res Commun* 324:849–854, 2004.
51. Weidemann T, Wachsmuth M, Tewes M, Rippe K and Langowski J. Analysis of ligand binding by two-colour fluorescence cross-correlation spectroscopy. *Single Mol* 3:49–61, 2002.
52. Heinze KG, Koltermann A and Schwille P. Simultaneous two-photon excitation of distinct labels for dual-color fluorescence crosscorrelation analysis. *Proc Natl Acad Sci USA* 97:10377–10382, 2000.
53. Denk W, Strickler JH and Webb WW. Two-photon laser scanning fluorescence microscopy. *Science* 248:73–76, 1990.
54. Xu C, Zipfel W, Shear JB, Williams RM and Webb WW. Multiphoton fluorescence excitation: new spectral windows for biological nonlinear microscopy. Proc of the Nat Acad of Sci USA 93:10763–10768, 1996.
55. Heinze KG, Koltermann A and Schwille P. Simultaneous two-photon excitation of distinct labels for dual-color fluorescence cross-correlation analysis. *Proc Nat Acad Sci USA* 97:10377–10382, 2000.
56. Le Grand Y, Leray A, Guilbert T and Odin C. Non-descanned versus descanned epifluorescence collection in two-photon microscopy: experiments and Monte Carlo simulations. *Opt Commun* 281:5480–5486, 2008.
57. Dittrich PS and Schwille P. Photobleaching and stabilization of. fluorophores used for single-molecule analysis. with one-and two-photon excitation. *Appl Phys B* 73:829–837, 2001.
58. Petrasek Z and Schwille P. Photobleaching in two-photon scanning fluorescence correlation spectroscopy. *Chemphyschem* 9:147–158, 2008.
59. Berland K and Shen G. Excitation saturation in two-photon fluorescence correlation spectroscopy. *Appl Opt* 42:5566–5576, 2003.
60. Koppel D. Statistical accuracy in fluorescence correlation spectroscopy. *Phys Rev A* 10:1938–1945, 1974.

61. Hwang LC and Wohland T. Dual-color fluorescence cross-correlation spectroscopy using single laser wavelength excitation. *Chemphyschem* 5:549–551, 2004.
62. Liu P, Sudhaharan T, Koh RM, Hwang LC, Ahmed S, Maruyama IN and Wohland T. Investigation of the dimerization of proteins from the epidermal growth factor receptor family by single wavelength fluorescence cross-correlation spectroscopy. *Biophys J* 93:684–698, 2007.
63. Sudhaharan T, Liu P, Foo YH, Bu W, Lim KB, Wohland T and Ahmed S. Determination of *in vivo* dissociation constant, Kd, of CDC42-effector complexes in live mammalian cells using single wavelength fluorescence cross-correlation spectroscopy (SW-FCCS). *J Biol Chem* 284:21100, 2009.
64. Kogure T, Karasawa S, Araki T, Saito K, Kinjo M and Miyawaki A. Fluorescent variant of a protein from the stony coral Montipora facilitates dual-color single-laser fluorescence cross-correlation spectroscopy. *Nat Biotechnol* 24:577–581, 2006.
65. Thews E, Gerken M, Eckert R, Zapfel J, Tietz C and Wrachtrup J. Cross talk free fluorescence cross correlation spectroscopy in live cells. *Biophys J* 89:2069–2076, 2005.
66. Dertinger T, Pacheco V, von der Hocht I., Hartmann R, Gregor I and Enderlein J. Two-focus fluorescence correlation spectroscopy: a new tool for accurate and absolute diffusion measurements. *Chemphyschem* 8:433–443, 2007.
67. Jaffiol R, Blancquaert Y, Delon A and Derouard J. Spatial fluorescence cross-correlation spectroscopy. *Appl Opt* 45:1225–1235, 2006.
68. Ries J and Schwille P. Studying slow membrane dynamics with continuous wave scanning fluorescence correlation spectroscopy. *Biophys J,* 91:1915–1924, 2006.
69. Beis D and Stainier DY. *In vivo* cell biology: following the zebrafish trend. *Trends Cell Biol* 16:105–112, 2006.
70. Korzh S, Pan X, Garcia-Lecea M., Winata CL, Wohland T, Korzh V and Gong Z. Requirement of vasculogenesis and blood circulation in late stages of liver growth in zebrafish. *BMC Dev Biol* 8:84, 2008.
71. Pan X, Shi X, Korzh V, Yu H and Wohland T. Line scan fluorescence correlation spectroscopy for three-dimensional microfluidic flow velocity measurements. *J Biome Opt* 14:024049, 2009.
72. Pan X, Yu H, Shi X, Korzh V and Wohland T. Characterization of flow direction in microchannels and zebrafish blood vessels by

scanning fluorescence correlation spectroscopy. *J Biome Opt* 12:014034, 2007.

73. Ries J, Yu SR, Burkhardt M, Brand M and Schwille P. Modular scanning FCS quantifies receptor-ligand interactions in living multicellular organisms. *Nat Methods* 6:643–645, 2009.

74. Shi X, Teo LS, Pan X, Chong SW, Kraut R, Korzh V and Wohland T. Probing events with single molecule sensitivity in zebrafish and Drosophila embryos by fluorescence correlation spectroscopy. *Dev Dyn* 238:3156–3167, 2009.

75. Lucitti JL, Jones EA, Huang C, Chen J, Fraser SE and Dickinson ME. Vascular remodeling of the mouse yolk sac requires hemodynamic force. *Development* 134:3317–3326, 2007.

76. Magde D, Webb WW and Elson EL. Fluorescence correlation spectroscopy. III. Uniform translation and laminar flow. *Biopolymers* 17:361–376, 1978.

77. Liang L, Wang X, Xing D, Chen T and Chen WR. Noninvasive determination of cell nucleoplasmic viscosity by fluorescence correlation spectroscopy. *J Biome Opt* 14:024013, 2009.

78. Pack C, Saito K, Tamura M and Kinjo M. Microenvironment and effect of energy depletion in the nucleus analyzed by mobility of multiple oligomeric EGFPs. *Biophys J* 91:3921–3936, 2006.

79. Higashijima S, Hotta Y and Okamoto H. Visualization of cranial motor neurons in live transgenic zebrafish expressing green fluorescent protein under the control of the islet-1 promoter/enhancer. *J Neurosci* 20:206–218, 2000.

80. Chong SW, Emelyanov A, Gong Z and Korzh V. Expression pattern of two zebrafish genes, cxcr4a and cxcr4b. *Mech Dev* 109:347–354, 2001.

81. Gilmour D, Knaut H, Maischein HM and Nusslein-Volhard C. Towing of sensory axons by their migrating target cells *in vivo*. *Nat Neurosci* 7:491–492, 2004.

82. Doitsidou M, Reichman-Fried M., Stebler J, Koprunner M, Dorries J, Meyer D, Esguerra CV, Leung T and Raz E. Guidance of primordial germ cell migration by the chemokine SDF-1. *Cell* 111:647–659, 2002.

83. Milon S, Hovius R, Vogel H and Wohland T. Factors influencing fluorescence correlation spectroscopy measurements on membranes: simulations and experiments. *Chem Phys* 288:171–186, 2003.

84. Barak LS, Ferguson SS, Zhang J, Martenson C, Meyer T and Caron MG. Internal trafficking and surface mobility of a functionally intact

beta2-adrenergic receptor-green fluorescent protein conjugate. *Mol Pharmacol* 51:177–184, 1997.

85. Babcock GJ, Farzan M and Sodroski J. Ligand-independent dimerization of CXCR4, a principal HIV-1 coreceptor. *J Biol Chem* 278:3378–3385, 2003.

86. Levoye A, Balabanian K, Baleux F, Bachelerie F and Lagane B. CXCR7 heterodimerizes with CXCR4 and regulates CXCL12-mediated G protein signalling. *Blood* 113:6085–6093, 2009.

87. Hart MJ, Callow MG, Souza B and Polakis P. IQGAP1, a calmodulin-binding protein with a rasGAP-related domain, is a potential effector for cdc42Hs. *EMBO J* 15:2997–3005, 1996.

88. Watanabe T, Wang S, Noritake J, Sato K, Fukata M, Takefuji M, Nakagawa M, Izumi N, Akiyama T and Kaibuchi K. Interaction with IQGAP1 links APC to Rac1, Cdc42, and actin filaments during cell polarization and migration. *Dev Cell* 7:871–883, 2004.

89. Ho YD, Joyal JL, Li Z and Sacks DB. IQGAP1 integrates Ca2+/calmodulin and Cdc42 signaling. *J Biol Chem* 274:464–470, 1999.

90. Mooney D, Hansen L, Vacanti J, Langer R, Farmer S and Ingber D. Switching from differentiation to growth in hepatocytes: control by extracellular matrix. *J Cell Physiol* 151:497–505, 1992.

91. Nagao I, Aoki Y, Tanaka M and Kinjo M. Analysis of the molecular dynamics of medaka nuage proteins by fluorescence correlation spectroscopy and fluorescence recovery after photobleaching. *FEBS J* 275:341–349, 2008.

92. Petrasek Z, Hoege C, Hyman A and Schwille P. Two-photon fluorescence imaging and correlation analysis applied to protein dynamics in *C. elegans* embryo. *Proc SPIE* 6860:68601L, 2008.

Chapter 5

# Real-Time Imaging of Lipid Metabolism in Larval Zebrafish

Juliana D. Carten and Steven A. Farber

*Department of Embryology, Carnegie Institution, Baltimore, MD USA*

## Abstract

Many fundamental questions remain regarding the cellular and molecular mechanisms of lipid metabolism. One major impediment to answering important questions in the field has been the lack of a tractable and sufficiently complex model system. Until recently, most studies of lipid metabolism have been performed *in vitro* or in the mouse, yet each approach possesses certain limitations. The zebrafish (*Danio rerio*) offers an excellent model system with which to study lipid metabolism *in vivo* due to its small size, genetic tractability and optical clarity. We have exploited the unique advantages of the zebrafish to visualize digestive processes *in vivo* by using a number of fluorescent tools, including fluorescent reporters of lipase and protease activity, fluorescent lipid analogs, and fluorescent microspheres. Using these tools with the zebrafish model system enables one to generate visible readouts of digestive physiology and organ function in real time. Additionally, the zebrafish system is amenable to high-throughput approaches to identify small molecules that influence lipid metabolism and new pharmaceuticals for the treatment of human lipid disorders. In this chapter we present recent advances in visualizing lipid metabolism in live larval zebrafish with a focus on fatty acid metabolism.

*Keywords*: Fatty Acids; Lipid; Triacylglyceride; Cholesterol; Satiety; Yolk; Larvae; Intestine.

Correspondence to: Dr. Steven A. Farber, Carnegie Institution for Science, 3520 San Martin Drive, Baltimore, MD 21218, USA. E-mail: farber@ciwemb.edu

# Introduction

The awarding of the 2008 Nobel Prize in chemistry for the discovery and development of green fluorescent protein (GFP),[1] highlights the importance of the ability to visualize biological processes. Previously invisible phenomena can now be observed using fluorescently tagged proteins[2] and fluorescent molecular analogs. From tracking the cell fates of stem cell progenitors[3] to the transport of cargo along axons in peripheral neurons,[4] we are now able to readily examine a variety of biological events in live cells. Most importantly, these fluorescent technologies have enabled the study of cellular events within the organs of live animals, providing insights into the regulation of physiological processes.[5,6]

# The Clinical Importance of Lipid Therapeutics

Lipid related diseases and their associated disorders are prevalent in many societies today. More than one-third of adults and 17% of children are currently classified as obese in the United States,[34,35] with obesity on the rise in many developing countries.[36] Defective lipid processing also underlies numerous chronic health conditions including cardiovascular disease, diabetes, fatty liver disease, dementia, and various metabolic syndromes. Thus, developing effective treatment strategies for these dislipidemias is of paramount importance, with potential worldwide impact.[37–39]

Only a handful of pharmaceuticals targeting lipid absorption have been made available in recent years, most notably the pancreatic lipase inhibitor Orlistat (Alli, GlaxoSmithKline) and the cholesterol absorption inhibitor Ezetimibe (Zetia, Merck/Schering Plough). The development of effective therapeutics for lipid disorders is largely hindered by gaps in our understanding of the basic molecular mechanisms underlying lipid transport and processing within cells. For example, exactly how cholesterol enters an absorptive cell in the intestine and the proteins that mediate this process are poorly understood.[40] Discovering small molecules for the treatment of human lipid disorders would be greatly facilitated by identifying key molecular targets within the physiological context of a live organism. Recent studies in the zebrafish have begun to utilize fluorescent lipid analogs and reporters to visualize lipid-processing events in optically clear larvae [5 to 7 days post-fertilization (dpf).[6,41–44] These techniques generate visible readouts of digestive physiology and allow whole animal and subcellular localization and assessment of lipid processing events. In addition, these approaches are amenable to scaling up for high throughput screens to identify drugs to treat lipid-associated diseases in a cost effective and time efficient manner.

## Limitations of *In Vitro* Lipid Metabolism Studies

There are a number of caveats that arise when attempting to study physiological processes regulated by complex chemical and neuronal cues *in vitro*. Studies designed to image lipid metabolism are often carried out in cells growing on artificial surfaces, such as plastic or glass, yet it is well known that adjacent tissues can greatly influence the behavior of cells. Additionally, cultured cells often lack many features of differentiated cells within an organ. Although the commonly used Caco-2 cell line exhibits intestinal cell-like features, it is derived from colorectal adenocarcinoma cells and has characteristics of transformed cells. Caco-2 cells also have variable villus structure[45] and altered levels of metabolic enzymes.[46] Furthermore, cell lines are often comprised of a single cell type and cannot duplicate the cellular heterogeneity of an organ, such as the intestine, that is composed of absorptive stem, enteroendocrine, immune, and goblet cells. The live intestine also contains symbiotic organisms, bile, and mucus that all influence lipid processing.[47–52] Due to these limitations of cultured cells, many long-standing questions in the lipid field, such as how a polarized intestinal epithelial cell absorbs and secretes dietary fat, remain unanswered. We advocate the use of the zebrafish model organism to conduct live studies of lipid metabolism. To this end we present a brief overview of emerging *in vivo* based strategies to acquire physiologically relevant information about lipid metabolism using larval zebrafish.

# Zebrafish as Models of Human Physiology and Disease

The zebrafish has proven to be an excellent model organism for studying a wide array of biological events. Initially used for embryological studies,[7–9] researchers have more recently exploited the zebrafish to visualize metabolism processes *in vivo*. Features of the zebrafish embryo that make it ideal for embryological studies (i.e. external fertilization, optical clarity, rapid development, and tractable genetics) also facilitate metabolic studies of organ function and physiology. The relative ease of use and cost effectiveness of zebrafish, as well as their high degree of genetic similarity to humans, has convinced many researchers that these fish can serve as disease models for a variety of human conditions. The zebrafish field abounds with models of human diseases, with zebrafish carrying a mutation in a specific disease gene manifesting phenotypes strikingly similar to those observed in humans.[10–28] Zebrafish models of human diseases include cardiovascular diseases (reviewed in Ref. 29), Duchenne muscular dystrophy,[30] various cancers (reviewed in Ref. 31), blood disorders,[32] obesity and others (only a small sampling from the last few years is cited above).[33]

# Lipid Metabolism is Conserved in Zebrafish

Six days after fertilization, larval zebrafish have developed many of the same gastrointestinal organs that are present in humans (i.e. the liver, intestinal bulb, pancreas, and gallbladder) with the exception of the stomach.[41,53–56] These organs are formed using similar genetic programs as in mammals.[41] The cellular composition of zebrafish digestive organs is also similar to mammals. The zebrafish intestine is composed of enterocytes, goblet and enteroendocrine cells.[51] Larvae also contain fat storing adipocytes, liver hepatocytes, and insulin-producing beta cells.[51,57] These cells, and the organs they comprise, are all visible during zebrafish development, facilitating the study of genes and small molecules that influence organ development and morphology.

As zebrafish embryos develop, yolk-derived lipids supply the energy and raw materials required for cellular divisions and tissue building. Yolk lipids are the source of cholesterol and fatty acids, which are needed to make cell membranes and bile acids.[58–61] Once larvae begin to eat, the intestinal epithelium takes on the function of metabolizing exogenous lipids to enable their transport to larval tissues. Like humans, zebrafish consume significant amounts of dietary lipids (*e.g.* triacylglycerol, fatty acids, phospholipids and cholesterol) and utilize similar lipid transport and lipolysis pathways.[62] Although many tissues can synthesize long-chain fatty acids and cholesterol endogenously, a zebrafish's yolk-sac and diet are its sources of essential lipids, which are critical for proper retinal and brain development.[63] The high degree of biochemical and functional homology between the larval zebrafish and human lipid metabolism clearly validates this system as a model for human lipid disorders.

## Dietary Lipid Metabolism in Zebrafish

Similarly to mammals, zebrafish consume fat in the form of triacylglycerol which must first be broken down into free fatty acids before being transported into the absorptive cells, termed enterocytes, of the gut.[64] Electron micrographs of zebrafish enterocytes show microvilli protruding into the intestinal lumen (Fig. 1). It is this apical region of microvilli, termed the brush border (BB), where fatty acid and cholesterol transport occurs across the plasma membrane. In fish and mammals dietary phospholipids and triacylglycerol are cleaved by intestinal and pancreatic lipases to form free fatty acids, lysolipid and phosphoglycerol.[65] These molecules are absorbed by enterocytes, presumably by simple diffusion (lysolipid, phosphoglycerol) and receptor-mediated processes.[66] In mammals absorbed fatty acids (FA) are processed according to their size: long chain fatty acids are re-esterified to form triglycerides and phospholipids, packaged into lipid drops, and then made into lipoproteins particles (chylomicrons or very low density lipoproteins (VLDL)) that are shuttled into the circulatory system via the lymphatics; shorter chain fatty acids enter the circulation directly via the portal vein due to their increased solubility.[67,68] Plasma lipoprotein-bound phospholipids and fatty acids subsequently enter peripheral tissues

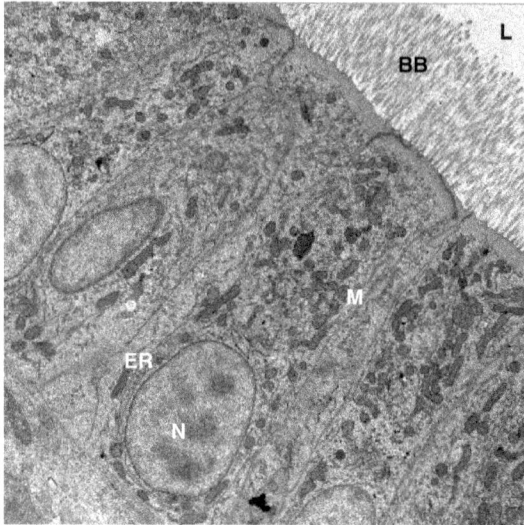

**Fig. 1** Electron micrograph of the zebrafish enterocyte. Subcellular structures of the enterocytes are labeled as nucleus (N), endoplasmic reticulum (ER), brush border (BB), lumen (L) and mitochondria (M). Image obtained from Dr. James Walters and printed with permission (data unpublished).

by binding to specific cell surface receptors or via endocytosis.[66,67] Recent work from our lab suggests that the metabolic steps which long and short chain fatty acids undergo in intestinal enterocytes are highly conserved in larval zebrafish (data unpublished).

An additional source of dietary fat is cholesterol, which mammals consume primarily in the form of cholesterol, and only 10%–15% as cholesteryl esters.[57,61] Cholesteryl esters are hydrolyzed by a specific pancreatic esterase to produce cholesterol and fatty acid. These hydrolyzed lipids form small lipid micelles with bile salts and are taken up by the intestine. Two endogenous sources of mammalian intestinal cholesterol are biliary cholesterol and the cholesterol contained in the membranes of dead epithelial cells that are sloughed off.[69] Although it is generally thought that cholesterol absorption occurs by a passive diffusion mechanism, recent studies have shown that cholesterol uptake by the intestine may occur via a protein-mediated process.[70–73]

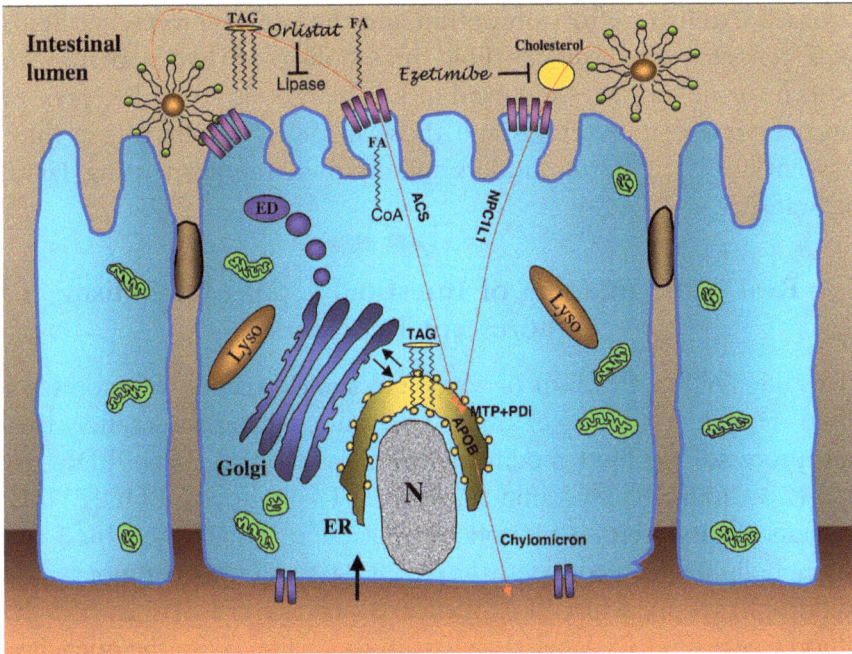

**Fig. 2** Vertebrate intestinal enterocyte transport. In the intestine, triacylglycerol (TAG), and phospholipids are hydrolyzed into fatty acids (FA) by lipases and in the intestinal lumen. FAs are presumably transported across the apical membrane of the enterocyte by FA transport proteins and then conjugated to Coenzyme A by acyl coenzyme A synthase (ACS). Acyl CoAs are used by various acyl transferases (not shown) to build TAG and cholesterol ester. In the lumen of the endoplasmic reticulum (ER), microsomal triacylglycerol transfer protein (MTP) and its obligate binding partner protein disulfide isomerase (Pdi) combine neutral lipids with ApoB to make chylomicrons. Chylomicrons are secreted across the basolateral membrane. Orlistat inhibits TAG hydrolysis and ezetimibe blocks cholesterol uptake possibly through the Niemann-Pick C1-Like (NPC1L1) transporter.

After cholesterol is taken-up by the mammalian intestine, it is re-esterified as cholesteryl ester by acyl Co-A: cholesterol acyltransferase (ACAT) and packaged into chylomicrons to be circulated through the lymphatic system (Fig. 2) and taken-up by peripheral tissues.

The high degree of biochemical and functional homology between zebrafish and mammalian lipid metabolism, validates this organism for the study of lipid metabolism. We have therefore focused extensive effort on optimizing the use of fluorescent lipids and reporters to visualize lipid metabolism in live zebrafish to screen for novel genes involved in key metabolic steps. We present an overview of our efforts and those made by others to use fluorescent tools to better understand lipid metabolism using larval zebrafish.

## Real-Time Imaging of Intestinal Lipid Metabolism: Fluorescent Reporters

To study lipid metabolism *in vivo*, we designed a family of fluorescent lipid reporters that alter their spectral characteristics after they are processed by lipid modifying enzymes[74,75] (Fig. 3A). PED6 [N-((6-(2,4-dinitro-phenyl)amino)hexanoyl)-1-palmitoyl-2-BODIPY-FL-pentanoyl-sn-glycerol-3-phosphoethanolamine] (Invitrogen Inc.) is a phospholipid with a covalently linked BODIPY fluorescent moiety and dinitrophenyl quencher. PLA$_2$-mediated cleavage of PED6 enables the fluorescent BODIPY-labeled acyl-chain to separate from the quencher, resulting in a significant increase in fluorescence.[74] After PED6

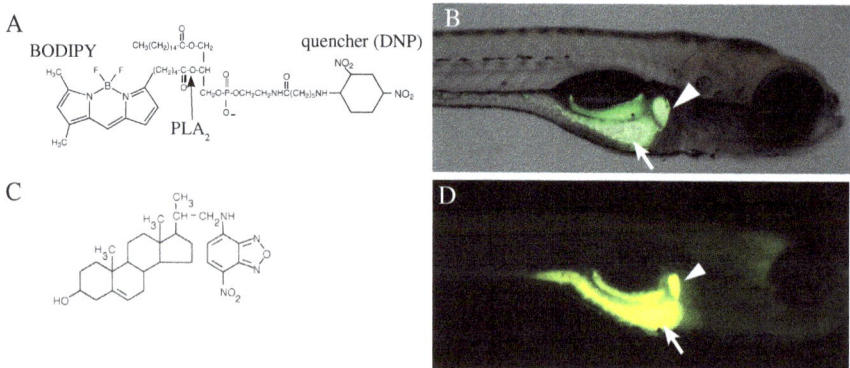

**Fig. 3** Fluorescent reporters as optical biosensors. Larvae soaked in PED6 (**A**) results in a fluorescent gallbladder and intestine (**B**). The same pattern of fluorescence is observed when larvae are soaked in unquenched NBD-cholesterol (**C** and **D**). Arrowhead marks gall bladder and arrow marks intestine.

ingestion, lipid processing can be visualized in live zebrafish, revealing an intensely labeled intestine, gall bladder and duct (Fig. 3B). Similarly, we developed a strategy to label larvae with a poorly soluble fluorescent cholesterol analog (NBD-Cholesterol) (Fig. 3C) by mixing it with purified fish bile. Feeding this mixture to 6 dpf larvae resulted in brilliant labeling of the larval digestive tract (Fig. 3D). An important difference between PED6 and NBD-Cholesterol is that NBD-cholesterol is not quenched and thus is perpetually fluorescent.

The utility of this approach for pharmaceutical development was validated when we treated embryos simultaneously with fluorescent reporters and small molecule inhibitors (statins) of the rate-limiting step in the *de novo* cholesterol synthesis pathway.[76] Statins are commonly prescribed for the treatment of hyperlipidemia and block the enzyme hydroxymethylglutaryl-coenzyme A reductase (HMGCoAR).[77] PED6 labeling of the gallbladder was completely blocked by the addition of atorvastatin (Lipitor, Pfizer) to the embryo media.[41] These findings demonstrated that larval zebrafish can be employed to identify new drugs that alter lipid metabolism and further confirmed that zebrafish process lipids by means similar to mammals.

Due to the ability of these fluorescent reporters to provide rapid readouts of lipid metabolism and digestive organ morphology in zebrafish larvae, our lab and collaborators were able to use them to perform the first physiological genetic screen in the zebrafish.[41,78] The first mutant we characterized in detail, *fat-free (ffr)*, has a digestive system that appears morphologically normal yet has impaired digestive lipid processing[41,79] (Fig. 4). We subsequently determined that this previously unidentified gene has a role in a number of cellular processes including Golgi structure, protein sorting, and intestinal lipid absorption.[79] Now that PED6 is commercially available, other laboratories have since utilized the reagent to assay digestive function in zebrafish[42,44] and in a genetic screen designed to identify genes involved in digestive organ formation and bile synthesis in Medaka fish.[80]

One general problem in using any reagent requiring ingestion is that perturbations in normal larval development (*e.g.* lower jaw

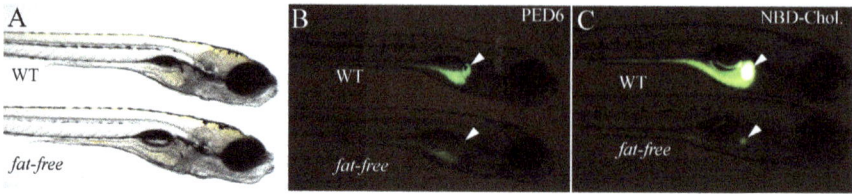

**Fig. 4** *fat-free* mutant larvae lack gallbladder fluorescence. (**A**) Bright field images of wild-type (WT) and mutant 5 dpf larvae. (**B**) Corresponding fluorescent images of the gallbladder following PED6 labeling (arrowheads). (**C**) *fat-free* larvae also fail to accumulate NBD-cholesterol.

formation, enteric neuron differentiation, and esophagus formation) result in attenuated fluorescence. Similarly, dead or sick larvae fail to label with a reagent like PED6 because swallowing is impaired. Thus, it is crucial to distinguish between primary and secondary lipid-associated defects. For example, when screening mutagenized families with PED6, a single pair cross of two heterozygous fish are expected to result in 25% of their progeny exhibiting severely reduced PED6 fluorescence in their gall bladders. While the failure of the gall bladder to become fluorescent may be due to a mutation that specifically alters lipid absorption or secretion (e.g. enterocyte chylomicron formation), it may also result from a dying embryo with a mutation affecting tRNA synthesis. In fact, the failure of larvae to utilize yolk and an underdeveloped jaw arise frequently when overall development has gone awry.[9] Therefore, additional secondary assays are required to check for viability and/or swallowing function if one is to identify genes or small molecules that specifically influence lipid metabolism.

## Fluorescent Microspheres

Small fluorescent microspheres are a useful way to assay swallowing function and intestinal viability of zebrafish larvae. These microspheres have broad biological applications and have been used in a number of developmental and toxicity studies. Bolcome and colleagues (2008) used microspheres to assess the structural integrity and permeability of zebrafish blood vessels in response to anthrax toxin, a poison known to cause human vascular dysfunction.[81] More recently, Field and others (2009) utilized microspheres to visualize peristaltic contractions within the intestine to determine if intestinal transit was defective in larvae with a compromised enteric nervous system.[5] Their studies found that intestinal transit significantly correlates with the degree of enteric innervation. Our lab has assayed swallowing function in *ffr* mutant larvae by placing small fluorescent microspheres in larval embryo media during fluorescent reporter soaking. We found that mutants swallowed the same number of beads as control larvae. This verified that decreased PED6 labeling in *ffr* mutants was caused by a primary lipid associated mutation and not a secondary, developmental abnormality that resulted in decreased swallowing ability (Fig. 5).[41]

## Triple Screening: PED6, EnzChek, and Microspheres

Although PED6 effectively labels digestive organs and can serve as a readout for organ function, there can be significant variation in reporter labeling between wild-type siblings (Fig. 6). This variation makes screening with PED6 difficult, as the expected 25% occurrence of recessive mutations in the progeny from a heterozygous cross may contain false positives. Furthermore, as previously mentioned, PED6 fluorescence is attenuated in dead or dying animals. To address the issue of variable labeling not caused by lipid-specific mutations, we developed an approach that utilized PED6 in conjunction with additional reporters of digestive function.[6] We first selected the commercially available reagent EnzChek (Invitrogen Inc.) to use as an additional digestive reporter. EnzChek is a fluorescently quenched phosphocasein protein that generates highly fluorescent casein fragments when proteolytically cleaved. Total fluorescence emitted from these fragments is proportional to enzyme activity.[82] EnzChek can therefore serve as a visible readout of protease activity (*e.g.* metallo, serine, acid and thiol proteases) in a number of biological systems.

A

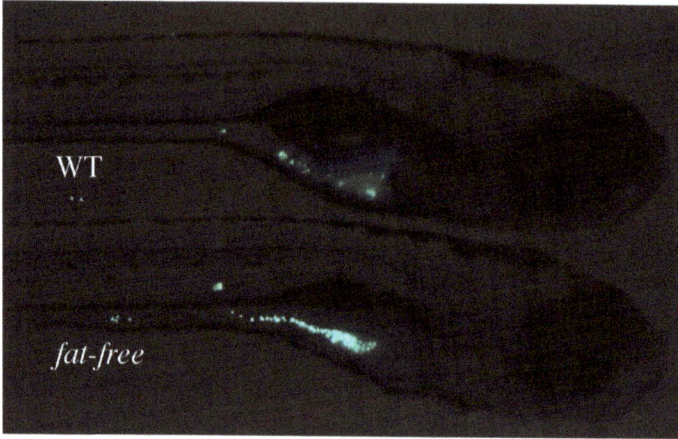

B

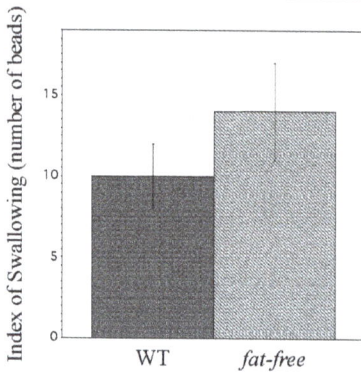

**Fig. 5** Swallowing is normal in the *fat-free* mutant. Larvae (6 dpf) were placed in embryo media containing fluorescent latex microspheres (0.0025% Fluoresbrite plain YG 2.0 $\mu$m, Polysciences Inc.) for 1 hour, washed, and imaged. Numbers of beads were 10± in the wild-type versus 14 ± 3 beads in *fat-free* mutant larvae (mean ± SEM, $n = 9$, $P > 0.3$). Figure reprinted from Carten and Farber (2009) with permission.[85]

We recently demonstrated that simultaneous feeding of PED6, the protease reporter EnzChek, and non-hydrolyzable microspheres allows one to monitor lipase, protease and swallowing activities, respectively, in live larval zebrafish.[6] Each reagent in the triple screening cocktail has a sufficiently distinct fluorescent emission, allowing the simultaneous viewing of all three signals (Fig. 7A). Labeling with the triple screen cocktail revealed a strong correlation between the intestinal protease and

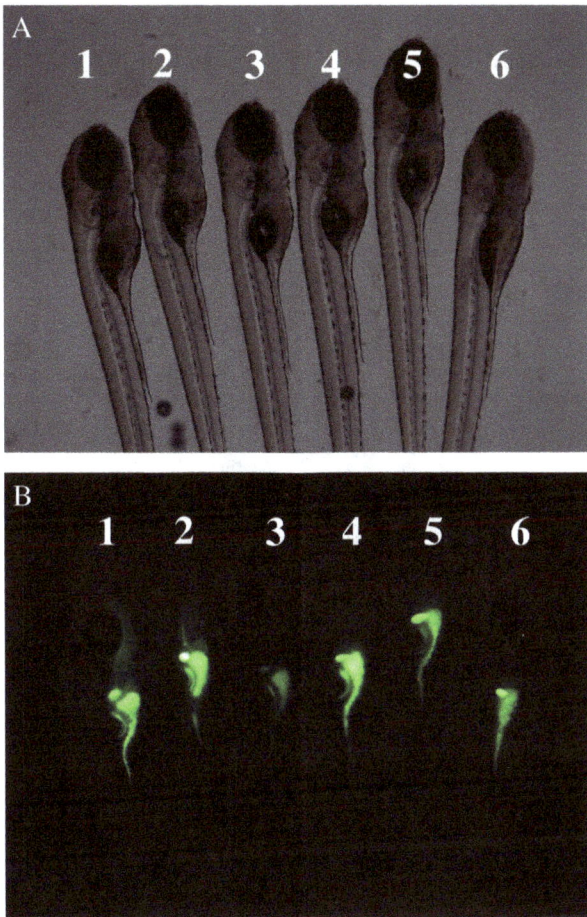

**Fig. 6** Wild-type siblings exhibit variability in PED6 signal. Larvae (6 dpf) were placed in embryo media containing PED6 (0.3 mg/ml) for 3 hours, washed, and imaged. (**A**) Bright field images of wild type 6 dpf larvae. (**B**) Corresponding fluorescent image.

lipase activity in wild-type larvae, suggesting that the variance observed in PED6 labeling was partly due to differing amounts of PED6 consumed by each larva (Figs. 7B and 7C). This work demonstrated that the ratio of PED6 to EnzChek fluorescence can serve as readout of digestive function, since the variability in the PED6/EnzChek ratio in multiple larvae is unaffected by differences in reporter ingestion.

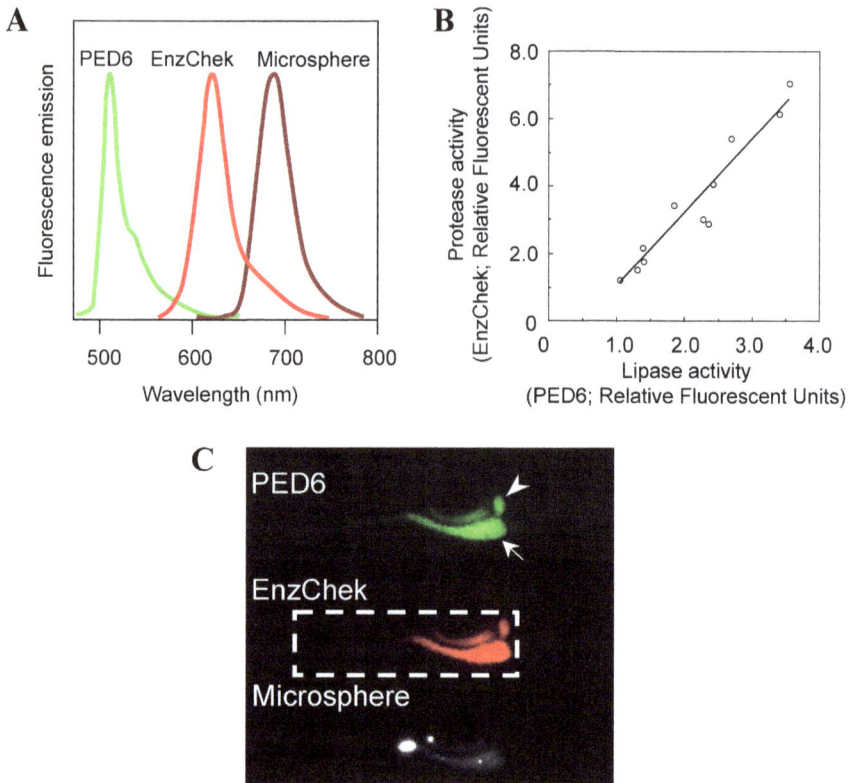

**Fig. 7** The triple screen assays lipase, protease and swallowing activity simultaneously. (**A**) PED6, EnzChek and microspheres each fluoresce at distinct wavelengths, allowing simultaneous screening for lipase, protease and swallowing activities, respectively. (**B**) Intestinal protease and phospholipase activity correlate with one another, allowing the ratio of PED6 to EnzChek signal to serve as a readout of digestive function. (**C**) PED6 and EnzChek signal in the gall bladder and intestine of wild-type zebrafish. (Arrowhead marks the gall bladder, arrow marks the intestine.) Microspheres are present in the intestine following feeding and indicate normal swallowing activity. Figures reprinted from Hama *et al.* (2008) with permission.[26]

After validating this approach, we used the triple screen to assay the function of the exocrine pancreas in larval zebrafish during development. Gastric lipases and proteases needed for the breakdown and subsequent uptake of nutrients are secreted by the exocrine pancreas.[83] Previous work has shown that morpholino knockdown of the *ptf1a* transcription factor can selectively prevent exocrine pancreas development.[57] Analysis of *ptf1a* mutants (5 dpf) using the triple screen found that these larvae retain normal levels of lipase activity yet have severely reduced protease activity. Older larvae (6 dpf) exhibit decreased amounts of both protease and lipase activity, suggesting the exocrine pancreas provides gastric lipases at later stages of development but not earlier. These data not only provides one more example of how digestive physiology previously described in mammals is similar to that observed in zebrafish, it also explains why dietary lipid processing studies in zebrafish performed before 6 dpf exhibit more variability.

The utility of the triple screening method was further demonstrated in our studies of the hormonal regulation of digestive enzyme activity. The peptide hormone cholecystokinin (CCK) facilitates digestion by causing the secretion of gastric enzymes into the small intestine after food consumption.[84] Release of CCK into the circulatory system activates the CCK receptor A (CCK-RA) in the exocrine pancreas. Larvae (5 dpf) treated with CCK-RA antagonist showed a reduction in protease activity but had unaffected lipase activity. Much like the *ptf1a* mutants, 6 dpf larvae had lower levels of both protease and lipase activity. Not surprisingly, the effect of CCK-RA antagonist was abolished in *ptf1a* morphants. This work suggests that CCK signaling regulates zebrafish secretion of exocrine pancreas-derived intestinal proteases earlier (5 dpf) and phospholipase activity later (6 dpf) in development. Taken together these data demonstrate that zebrafish digestive physiology is very similar to that of mammals in that the exocrine pancreas secretes key lipid processing enzymes and that this process is hormonally controlled. More importantly, it demonstrates that the zebrafish can be used to identify new small molecules that influence lipid metabolism.

## Concluding Thoughts

Despite the immense potential of zebrafish as a therapeutic screening tool, this model organism is currently underutilized by both the academic and pharmaceutical research communities. The zebrafish has been primarily utilized to study embryologic questions specifically focused on early patterning; however, more researchers are discovering the system for studies of physiology and disease. This trend is likely to continue with increasing numbers of groups exploiting the tractable genetics of the larval zebrafish to discover new genes that regulate important physiological processes. These efforts are bolstered by the ease with which human genes can be identified from zebrafish orthologs by searching publicly available databases. Furthermore, zebrafish larvae possess a number of advantages for drug development. Most notably the larvae exhibit the molecular complexity found in higher vertebrates and thus are ideally suited for new drug target iden-tification. Zebrafish larvae also readily absorb compounds from the water directly into their circulatory system without permeabilization. Despite these advantages, little attention has been focused on the study of metabolic processes in this organism. While pharmaceutical compa-nies are beginning to consider the zebrafish model system, most ongoing high-throughput small molecule screens assay for inhibitors of single molecule-interactions. We advocate whole animal-based screens for drug discovery and expect to see the zebrafish system utilized more frequently for these efforts. In many animals certain compounds are metabolized into new compounds that may be significantly more bioac-tive than the starting compound (i.e. the intestinal cholesterol absorption inhibitor ezetimibe). It is entirely possible that compounds used in conventional screens are found to be inactive because they are not modified by host enzymes. A whole animal screening approach in zebrafish would potentially identify new bioactive compounds that will be critical for addressing the growing list of disorders and diseases asso-ciated with abnormal lipid metabolism.

## Acknowledgments

The authors would like to thank Jennifer Anderson for editorial assis-tance. We also thank James Walters and Mike Sepanski for the electron micrograph image of the zebrafish enterocyte.

The description of the triple screen above was used, with permission, from a review previously published by Carten and Farber (2009).[85]

# Financial Disclosure

The authors have no affiliations or financial arrangement with any organization that has a financial interest or stake in the material discussed in this manuscript. The Carnegie Institution does hold a patent together with the University of Penn sylvania on the invention of the author (S.A.F.) that describes the use of fluorescent lipids in zebrafish for high-throughput screening Pub. No.: US 2009/0136428 A1. The authors have no current consultancies, honoraria, stock ownership or options, expert testimony or royalties regarding the material described. However, the Carnegie Institution and/or U. Penn. can license this technology in the future potentially providing royalties to the author (S.A.F). Research performed by the authors and described in this manuscript was supported by the Carnegie Institution Endowment and with grants from the US National Institutes of Health (RO1 GM63904 and RO1 DK060369).

# References

1. The Nobel Foundation http://nobelprize.org/nobel_prizes/chemistry/laureates/2008/.
2. Detrich HW, 3rd. Fluorescent proteins in zebrafish cell and developmental biology. *Methods Cell Biol* 85:219–241, 2008.
3. Zernicka-Goetz M, Pines J, McLean Hunter S, Dixon JP, Siemering KR, Haseloff J and Evans MJ. Following cell fate in the living mouse embryo. *Development* 124:1133–1137, 1997.
4. Prekeris R, Foletti DL and Scheller RH. Dynamics of tubulovesicular recycling endosomes in hippocampal neurons. *J Neurosci* 19:10324–10337, 1999.
5. Field HA, Kelley KA, Martell L, Goldstein AM and Serluca FC. Analysis of gastrointestinal physiology using a novel intestinal transit assay in zebrafish *Neurogastroenterol Motil* 21:304–312, 2009.
6. Hama K, Provost E, Baranowski TC, Rubinstein AL, Anderson JL, Leach SD and SA F. *In vivo* imaging of zebrafish digestive organ function using multiple quenched fluorescent reporters. *Am J Physiol Gastrointest Liver Physiol* 296:G445–G453, 2009.

7. Jones RW and Huffman MN. Fish embryos as bioassay material in testing chemicals for effect on cell division and differentiation. *Trans Am Micro sc Soc* 76:177–183, 1957.

8. Kimmel CB, Sessions SK and Kimmel RJ. Morphogenesis and synaptogenesis of the zebrafish Mauthner neuron. *J Comp Neurol* 198:101–120, 1981.

9. Haffter P, Granato M, Brand M, Mullins MC, Hammerschmidt M, Kane DA, Odenthal J, van Eeden FJ, Jiang YJ, Heisenberg CP, Kelsh RN, Furutani-Seiki M, Vogelsang E, Beuchle D , Schach U, Fabian C and Nusslein-Volhard C. The identification of genes with unique and essential functions in the development of the zebrafish, *Danio rerio*. *Development* 123:1–36, 1996.

10. Amali AA, Rekha RD, Lin CJ, Wang WL, Gong HY, Her GM and Wu JL. Thioacetamide induced liver damage in zebrafish embryo as a disease model for steatohepatitis. *J Biomed Sci* 13:225–232, 2006.

11. Bai Q, Mullett SJ, Garver JA, Hinkle DA and Burton EA. Zebrafish DJ-1 is evolutionarily conserved and expressed in dopaminergic neurons. *Brain Res* 1113:33–44, 2006.

12. Berghmans S, Murphey RD, Wienholds E, Neuberg D, Kutok JL, Fletcher CD, Morris JP, Liu TX, Schulte-Merker S, Kanki JP, Plasterk R, Zon LI and Look AT. tp53 mutant zebrafish develop malignant peripheral nerve sheath tumors. *Proc Natl Acad Sci USA* 102:407–412, 2005.

13. Fetcho JR. The utility of zebrafish for studies of the comparative biology of motor systems. *J Exp Zool B Mol Dev Evol* 308:550–562, 2007.

14. Fisher S, Jagadeeswaran P and Halpern ME. Radiographic analysis of zebrafish skeletal defects. *Dev Biol* 264:64–76, 2003.

15. Guyon JR, Steffen LS, Howell MH, Pusack TJ, Lawrence C and Kunkel LM. Modeling human muscle disease in zebrafish. *Biochim Biophys Acta* 1772:205–215, 2007.

16. Khuchua Z, Yue Z, Batts L and Strauss AW. A zebrafish model of human Barth syndrome reveals the essential role of tafazzin in cardiac development and function. *Circ Res* 99:201–208, 2006.

17. Kunkel LM, Bachrach E, Bennett RR, Guyon J and Steffen L. Diagnosis and cell-based therapy for Duchenne muscular dystrophy in humans, mice, and zebrafish. *J Hum Genet* 51:397–406, 2006.

18. Lam SH and Gong Z. Modeling liver cancer using zebrafish: a comparative oncogenomics approach. *Cell Cycle* 5:573–577, 2006.

19. Lam SH, Wu YL, Vega VB, Miller LD, Spitsbergen J, Tong Y, Zhan H, Govindarajan KR, Lee S, Mathavan S, Murthy KR, Buhler DR, Liu ET and Gong Z. Conservation of gene expression signatures between zebrafish and human liver tumors and tumor progression. *Nat Biotechnol* 24:73–75, 2006.
20. Langenau DM and Zon LI. The zebrafish: a new model of T-cell and thymic development. *Nat Rev Immunol* 5:307–317, 2005.
21. Langheinrich U, Vacun G and Wagner T. Zebrafish embryos express an orthologue of HERG and are sensitive toward a range of QT-prolonging drugs inducing severe arrhythmia. *Toxicol Appl Pharmacol* 193:370–382, 2003.
22. McKinley ET, Baranowski TC, Blavo DO, Cato C, Doan TN and Rubinstein AL. Neuroprotection of MPTP-induced toxicity in zebrafish dopaminergic neurons. *Brain Res Mol Brain Res* 141:128–137, 2005.
23. Norrby K. *In vivo* models of angiogenesis. *J Cell Mol Med* 10:588–612, 2006.
24. Obara T, Mangos S, Liu Y, Zhao J, Wiessner S, Kramer-Zucker AG, Olale F, Schier AF and Drummond IA. Polycystin-2 immunolocalization and function in zebrafish. *J Am Soc Nephrol* 17:2706–2718, 2006.
25. Sabaawy HE, Azuma M, Embree LJ, Tsai HJ, Starost MF and Hickstein DD. TEL-AML1 transgenic zebrafish model of precursor B cell acute lymphoblastic leukemia. *Proc Natl Acad Sci USA* 103:15166–15171, 2006.
26. Sayer JA, Otto EA, O'Toole JF, Nurnberg G, Kennedy MA, Becker C, Hennies HC, Helou J, Attanasio M, Fausett BV, Utsch B, Khanna H, Liu Y, Drummond I, Kawakami I, Kusakabe T, Tsuda M, Ma L, Lee H, Larson RG, Allen SJ, Wilkinson CJ, Nigg EA, Shou C, Lillo C, Williams DS, Hoppe B, Kemper MJ, Neuhaus T, Parisi MA, Glass IA, Petry M, Kispert A, Gloy J, Ganner A, Walz G, Zhu X, Goldman D, Nurnberg P, Swaroop A, Leroux MR and Hildebrandt F. The centrosomal protein nephrocystin-6 is mutated in Joubert syndrome and activates transcription factor ATF4. *Nat Genet* 38:674–681, 2006.
27. Titus TA, Selvig DR, Qin B, Wilson C, Starks AM, Roe BA and Postlethwait JH. The Fanconi anemia gene network is conserved from zebrafish to human. *Gene* 371:211–223, 2006.
28. Williamson KA, Hever AM, Rainger J, Rogers RC, Magee A, Fiedler Z, Keng WT, Sharkey FH, McGill N, Hill CJ, Schneider A, Messina M, Turnpenny PD, Fantes JA, van Heyningen V and FitzPatrick DR.

Mutations in SOX2 cause anophthalmia-esophageal-genital (AEG) syndrome. *Hum Mol Genet* 15:1413–1422, 2006.

29. Chico TJA, Ingham PW and Crossman DC. Modeling cardiovascular disease in the zebrafish. *Trends Cardiovasc Med* 18:150–155, 2008.

30. Parsons MJ, Campos I, Hirst EM and Stemple DL. Removal of dystroglycan causes severe muscular dystrophy in zebrafish embryos. *Development* 129:3505–3512, 2002.

31. Feitsma H and Cuppen E. Zebrafish as a cancer model. *Mol Cancer Res* 6:685-694, 2008.

32. Stainier DY, Weinstein BM, Detrich HW, 3rd, Zon LI and Fishman MC. Positional cloning of the zebrafish sauternes gene: a model for congenital sideroblastic anaemia [see comments]. *Nat Genet* 20:244–250, 1998.

33. Song Y and Cone RD. Creation of a genetic model of obesity in a teleost. *FASEB J* 21:2042–2049, 2007.

34. Ford ES and Mokdad AH. Epidemiology of obesity in the Western Hemisphere. *J Clin Endocrinol Metab* 93:S1–8, 2008.

35. Wang Y and Beydoun MA. The obesity epidemic in the United States — gender, age, socioeconomic, racial/ethnic, and geographic characteristics: a systematic review and meta-regression analysis. *Epidemiol Rev* 29:6–28, 2007.

36. Misra A and Khurana L. Obesity and the metabolic syndrome in developing countries. *J Clin Endocrinol Metab* 93:S9–30, 2008.

37. Joffe BI, Panz VR, and Raal FJ. From lipidodystrophy syndromes to diabetes mellitus. *Lancet* 357:1379–1381, 2001.

38. McNeely MJ, Edwards KL, Marcovina SM, Brunzell JD, Motulsky AG and Austin MA. Lipoprotein and alipoprotein abnormalities in familial combined hyperlipidemia: a 20 year prospective study. *Atherosclerosis* 159:471–481, 2001.

39. Watanabe S, Yaginuma R, Ikejima K and Miyazaki A. Liver diseases and metabolic syndrome. *J Gastroenterol* 43:509–518, 2008.

40. Klett EL and Patel SB. Biomedicine. Will the real cholesterol transporter please stand up. *Science* 303:1149–1150, 2004.

41. Farber SA, Pack M, Ho SY, Johnson ID, Wagner DS, Dosch R, Mullins MC, Hendrickson HS, Hendrickson EK and Halpern ME. Genetic analysis of digestive physiology using fluorescent phospholipid reporters. *Science* 292:1385–1388, 2001.

42. Sadler KC, Amsterdam A, Soroka C, Boyer J and Hopkins N. A genetic screen in zebrafish identifies the mutants vps18, nf2 and foie gras as models of liver disease. *Development* 132:3561–3572, 2005.

43. Schlegel A and Stainier DYR. Lessons from "lower" organisms: what worms, flies, and zebrafish can teach us about human energy metabolism. *PLoS Genet* 3:e199, 2007.
44. Schlegel A and Stainier DY. Microsomal triglyceride transfer protein is required for yolk lipid utilization and absorption of dietary lipids in zebrafish larvae. *Biochemistry* 45:15179–15187, 2006.
45. Rousset M. The human colon carcinoma cell lines HT-29 and Caco-2: two *in vitro* models for the study of intestinal differentiation. *Biochimie* 68:1035–1040, 1986.
46. Trotter PJ and Storch J. Fatty acid esterification during differentiation of the human intestinal cell line Caco-2. *J Biol Chem* 268:10017–10023, 1993.
47. Field HA, Ober EA, Roeser T and Stainier DY. Formation of the digestive system in zebrafish. I. Liver morphogenesis. *Dev Biol* 253:279–290, 2003.
48. Kruit JK, Groen AK, van Berkel TJ and Kuipers F. Emerging roles of the intestine in control of cholesterol metabolism. *World J Gastroenterol* 12:6429–6439, 2006.
49. Martin FP, Wang Y, Sprenger N, Yap IK, Lundstedt T, Lek P, Rezzi S, Ramadan Z, van Bladeren P, Fay LB, Kochhar S, Lindon JC, Holmes E and Nicholson JK. Probiotic modulation of symbiotic gut microbial-host metabolic interactions in a humanized microbiome mouse model. *Mol Syst Biol* 4:157, 2008.
50. Moschetta A, Xu F, Hagey LR, van Berge-Henegouwen GP, van Erpecum KJ, Brouwers JF, Cohen JC, Bierman M, Hobbs HH, Steinbach JH and Hofmann AF. A phylogenetic survey of biliary lipids in vertebrates. *J Lipid Res* 46:2221–2232, 2005.
51. Pack M, Solnica-Krezel L, Malicki J, Neuhauss SC, Schier AF, Stemple DL, Driever W and Fishman MC. Mutations affecting development of zebrafish digestive organs. *Development* 123:321–328, 1996.
52. Titus E and Ahearn GA. Vertebrate gastrointestinal fermentation: transport mechanisms for volatile fatty acids. *Am J Physiol* 262:R547–553, 1992.
53. Lieschke GJ and Currie PD. Animal models of human disease: zebrafish swim into view. *Nat Rev Genet* 8:353–367, 2007.
54. Wallace KN and Pack M. Unique and conserved aspects of gut development in zebrafish. *Dev Biol* 255:12–29, 2003.
55. Wallace KN, Yusuff S, Sonntag JM, Chin AJ and Pack M. Zebrafish hhex regulates liver development and digestive organ chirality. *Genesis* 30:141–143, 2001.

56. Yee NS, Yusuff S, and Pack M. Zebrafish pdx1 morphant displays defects in pancreas development and digestive organ chirality, and potentially identifies a multipotent pancreas progenitor cell. *Genesis* 30:137–140, 2001.

57. Lin J, Biankin AV, Horb ME, Ghosh B, Prasad NB, Yee NS, Pack MA and Leach SD. Differential requirement for ptf1a in endocrine and exocrine lineages of developing zebrafish pancreas. *Dev Biol* 274:491–503, 2004.

58. Babin PJ, Thisse C, Durliat M, Andre M, Akimenko MA and Thisse B. Both apolipoprotein E and A-I genes are present in a nonmammalian vertebrate and are highly expressed during embryonic development. *Proc Natl Acad Sci USA* 94:8622–8627, 1997.

59. Bownes M. Why is there sequence similarity between insect yolk proteins and vertebrate lipases? *J Lipid Res* 33:777–790, 1992.

60. Munoz G, Donghi S and Cerisola H. Vitellogenesis in the crayfish Rhynchocinetes typus: role of hepatopancreas in lipid yolk biosynthesis. *Cell Mol Biol* 36:531–536, 1990.

61. Angelin B, Eriksson M and Rudling M. Bile acids and lipoprotein metabolism: a renaissance for bile acids in the post-statin era? *Curr Opin Lipidol* 10:269–274, 1999.

62. Sheridan MA. Lipid dynamics in fish: aspects of absorption, transportation, deposition and mobilization. *Comp Biochem Physiol B* 90:679–690, 1988.

63. Uauy R and Dangour AD. Nutrition in brain development and aging: role of essential fatty acids. *Nutr Rev* 64:S24–33; discussion S72–91, 2006.

64. Tso P and Fujimoto K. The absorption and transport of lipids by the small intestine. *Brain Res Bull* 27:477–482, 1991.

65. Tocher D. Glycerophospholipid metabolism. In: *Biochemistry and Molecular Biology of Fishes.* Hochachka P and Mommsen T (eds.) Elsevier, New York, 1995, pp. 119–157.

66. Spector AA. Plasma lipid transport. *Clin Physiol Biochem* 2:123–134, 1984.

67. Glatz JF and van der Vusse GJ. Intracellular transport of lipids. *Mol Cell Biochem* 88:37–44, 1989.

68. Tso P, Liu M and Kalogeris TJ. The role of apolipoprotein A-IV in food intake regulation. *J Nutrition* 129:1503–1506, 1999.

69. Ros E. Intestinal absorption of triglyceride and cholesterol. Dietary and pharmacological inhibition to reduce cardiovascular risk [In Process Citation]. *Atherosclerosis* 151:357–379, 2000.

70. Altmann SW, Davis HR, Jr, Zhu LJ, Yao X, Hoos LM, Tetzloff G, Iyer SP, Maguire M, Golovko A, Zeng M, Wang L, Murgolo N and Graziano MP. Niemann-Pick C1 Like 1 protein is critical for intestinal cholesterol absorption. *Science* 303:1201–1204, 2004.

71. Davies JP, Scott C, Oishi K, Liapis A and Ioannou YA. Inactivation of NPC1L1 causes multiple lipid transport defects and protects against diet-induced hypercholesterolemia. *J Biol Chem* 280:12710–12720, 2005.

72. Hauser H, Dyer JH, Nandy A, Vega MA, Werder M, Bieliauskaite E, Weber FE, Compassi S, Gemperli A, Boffelli D, Wehrli E, Schulthess G and Phillips MC. Identification of a receptor mediating absorption of dietary cholesterol in the intestine. *Biochemistry* 37:17843–17850, 1998.

73. Thurnhofer H and Hauser H. Uptake of cholesterol by small intestinal brush border membrane is protein-mediated. *Biochemistry* 29:2142–2148, 1990.

74. Farber SA, Olson ES, Clark JD and Halpern ME. Characterization of $Ca^{2+}$-dependent phospholipase A2 activity during zebrafish embryogenesis. *J Biol Chem* 274:19338–19346, 1999.

75. Hendrickson HS, Hendrickson EK, Johnson ID and Farber SA. Intramolecularly quenched BODIPY-labeled phospholipid analogs in phospholipase A(2) and platelet-activating factor acetylhydrolase assays and *in vivo* fluorescence imaging. *Anal Biochem* 276:27–35, 1999.

76. Thorpe JL, Doitsidou M, Ho SY, Raz E and Farber SA. Germ cell migration in zebrafish is dependent on HMGCoA reductase activity and prenylation. *Dev Cell* 6:295–302, 2004.

77. Brown MS and Goldstein JL. The LDL receptor and HMG-CoA reductase — two membrane molecules that regulate cholesterol homeostasis. *Curr Top Cell Regul* 26:3–15, 1985.

78. Fishman MC. GENOMICS: Zebrafish — the canonical vertebrate. *Science* 294:1290–1291, 2001.

79. Ho SY, Lorent K, Pack M and Farber SA. Zebrafish fat-free is required for intestinal lipid absorption and Golgi apparatus structure. *Cell Metab* 3:289–300, 2006.

80. Watanabe T, Asaka S, Kitagawa D, Saito K, Kurashige R, Sasado T, Morinaga C, Suwa H, Niwa K, Henrich T, Hirose Y, Yasuoka A, Yodae H, Deguchi T, Iwanami N, Kunimatsu S, Osakada M, Loosli F, Quiring R, Carl M, Grabher C, Winkler S, Del Bene F, Wittbrodt J, Abe K, Takahama Y, Takahashi K, Katada T, Nishina H, Kondoh H and

Furutani- Seiki M. Mutations affecting liver development and function in Medaka, *Oryzias latipes,* screened by multiple criteria. *Mech Dev* 121:791–802, 2004.

81. Bolcome RE, Sullivan SE, Zeller R, Barker AP, Collier RJ and Chan J. Anthrax lethal toxin induces cell death-independent permeability in zebrafish vasculature. *Proc Natl Acad Sci* 105:2439–2444, 2008.

82. Jones LJ, Upson RH, Haugland RP, Panchuk-Voloshina N and Zhou M. Quenched BODIPY dye-labeled casein substrates for the assay of protease activity by direct fluorescence measurement. *Anal Biochem* 251:144–152, 1997.

83. Layer Pa and Keller J. Pancreatic enzymes: secretion and luminal nutrient digestion in health and disease. *J Clin Gastroenterol* 28:3-10, 1999.

84. Raybould HE. Mechanisms of CCK signaling from gut to brain. *Curr Opin Pharmacol* 7:570–574, 2007.

85. Carten JD and Farber SA. A new model system swims into focus: using the zebrafish to visualize intestinal lipid metabolism *in vivo*. *Clin Lipidol* 4:501–515, 2009.

# Chapter 6

# Live Imaging Innate Immune Cell Behavior During Normal Development, Wound Healing and Infection

Chris Hall, Maria Vega Flores, Makoto Kamei, Kathryn Crosier and Phil Crosier

*Department of Molecular Medicine and Pathology*
*School of Medical Sciences, The University of Auckland*
*Auckland, New Zealand*

## Abstract

Cells of the innate immune system perform a number for functions that range from clearing apoptotic cell corpses during normal organ development to removing invading pathogens and helping regulate the immune response to infection. Traditional approaches to assessment of these immune cell behaviors have relied upon histological analysis of fixed tissue samples complemented by *in vitro* functional data. Despite providing significant insights, translating the results from such studies into a multicellular whole animal context is difficult. Recent advances in cell labeling techniques and imaging technologies has given researchers unprecedented ability to directly observe immune cell activities within a live whole animal context. Such live imaging approaches are essential to completely appreciate these

Correspondence to: Dr. Phil Crosier, Department of Molecular Medicine and Pathology, Faculty of Medical and Health Sciences, The University of Auckland, 85 Park Road, Grafton, Private Bag 92019, Auckland, New Zealand, E-mail: ps.crosier@auckland.ac.nz activities within their intact normal physiological setting. The zebrafish embryo, with its optical transparency and fully functional innate and adaptive immune systems, represents an ideal system in which to live image innate immune cells. When this is coupled with the relative ease with which specific immune cells can be fluorescently labeled within transgenic reporter lines and the ability to both genetically and chemically interfere with genetic pathways, the zebrafish is emerging as a strong vertebrate platform in which to make novel contributions to understanding immune cell biology. This chapter will discuss how our group and others use live imaging to assess the behavior of myeloid leukocytes during normal development and in response to different inflammatory stimuli, including wounding and pathogenic challenge.

*Keywords*: Zebrafish; Leukocyte; Innate Immunity; Live Imaging; Wound Healing; Infection; Confocal Microscopy; Transgenic.

## Introduction

Live imaging of specific immune cell subsets is beginning to reveal novel insights into how these cells participate during inflammatory responses and how they interact with each other and various inflammatory triggers. A fundamental aspect of a robust immune cell response is the inherent capacity of cells comprising the innate immune compartment to migrate from disparate tissues within the host and converge on the site of infection/inflammation. An intimate understanding of this migration and the subsequent functions of these cells following infiltration is central to fully appreciating how the innate arm of the immune system helps to govern the magnitude and duration of an inflammatory response. Live imaging of the immune system in mammals has relied predominantly on two techniques, both of which require invasive surgical manipulations. The first, tissue explant assays, requires the surgical removal of the organ of interest while the second, intravital imaging, requires surgical exposure of the tissue(s) to be imaged.[1,2] Each technique presents its own set of technical issues, including the need to maintain important vascular, lymphatic and neural circuitry.[1,2] Although these imaging strategies enable observation of immune cells within various tissue contexts, the effect of surgical trauma on the behaviors of the immune cells being studied remains difficult to assess. The ultimate goal in live imaging of the immune system is to observe immune cells in real-time within their undisturbed multi-cellular environment. The unique combination of exquisite optical transparency, external development and small size positions the zebrafish as an ideal platform in which to realize such a whole animal approach to live imaging immune responses. Coupling these attributes with the genetic tractability of the zebrafish system provides a powerful means to genetically dissect the factors that regulate immune cell function.

Like all jawed vertebrates, zebrafish possess both innate and adaptive immune cell compartments that resemble those of mammals.[3-5] In addition, despite considerable evolutionary divergence the ontogeny of these cells and the genetic drivers of their differentiation remain remarkably conserved.[6] In zebrafish, an initial "primitive" wave of hematopoiesis equips the pre-circulation embryo with macrophages that colonize the yolk prior to entering the embryonic circulation.[7] A subset of these early myeloid cells also possess the potential to differentiate into neutrophils following invasion of embryonic tissues.[8] These primitive myeloid leukocytes are then reinforced by those derived from "definitive" erythromyeloid precursor cells that emerge *de novo* within the posterior blood island; an hematopoietic niche that develops in close association with the tail vasculature.[9] These myeloid leukocytes provide the embryo and larva with an innate cellular defense system prior to establishment of definitive hematopoiesis in the kidney, the equivalent of the mammalian bone marrow. Initial studies exploiting the optical transparency of zebrafish embryos to assess the immune potential of these cells used differential interference contrast (DIC) video-microscopy to observe the real-time response of superficially located macrophages in helping to eradicate bacterial challenges.[7] Further studies enhanced this approach by fluorescently labeling the bacterial trigger.[10,11] Our group and others have taken this a step further by fluorescently marking specific myeloid leukocyte compartments, providing a unique opportunity to observe how these innate immune cells behave within an intact whole vertebrate setting.[12-16] This chapter will discuss how we have exploited the physical traits of the zebrafish to generate a whole animal platform for live real-time imaging of myeloid leukocytes, to investigate their function during normal development and during inflammatory responses.

## Mounting Strategies for Live Imaging

We employ short-, medium- and long-term embryo embedding strategies that become more labor-intensive as the duration of imaging increases (Fig. 1). All imaging described in this chapter was performed using an Olympus FV1000 scanning confocal microscope equipped with long working distance 20X and 60X water immersion lenses and an incubation chamber. For time-lapse imaging less than four hours in duration, we simply embed the embryo/larva in a 35 mm plastic tissue culture dish. Initially an agarose bed (1% low melting point (lmp) agarose in E3 medium (w/v) supplemented with 0.003% 1-phenyl-2-thiourea (PTU) and 0.168 mg/ml tricaine) is laid and when solidified a small well is excavated to create space to accommodate the yolk when orientating the specimen. The anesthetized embryo to be imaged is then placed on the agarose bed and immediately covered with a thin layer (~1 mm) of molten 1% embedding agarose (1% lmp agarose in E3 supplemented with PTU and tricaine). While solidifying, the embryo/larva is manipulated into the correct orientation for imaging; for example, when imaging ventral fin wounds the yolk is positioned in the orientation well enabling the embryo/larva to lay flat on its lateral surface to create a horizontal imaging plane (Fig. 1A). Once solidified, the agarose is overlaid with E3 medium (similarly supplemented with PTU and tricaine). Despite being a simple approach to immobilize the specimen during imaging, for imaging extending beyond four hours the embryos/larvae begin to demonstrate signs of stress including progressively labored circulation, tissue necrosis and ultimately cardiac arrest and death. To maintain the embryos/larvae for imaging periods extending up to 18 hours, we modified the above technique (as previously described[17]). Briefly, instead of covering the entire specimen with agarose, a small amount is placed directly on the embryo and the embryo is positioned as above (Fig. 1B). Once solidified, the head

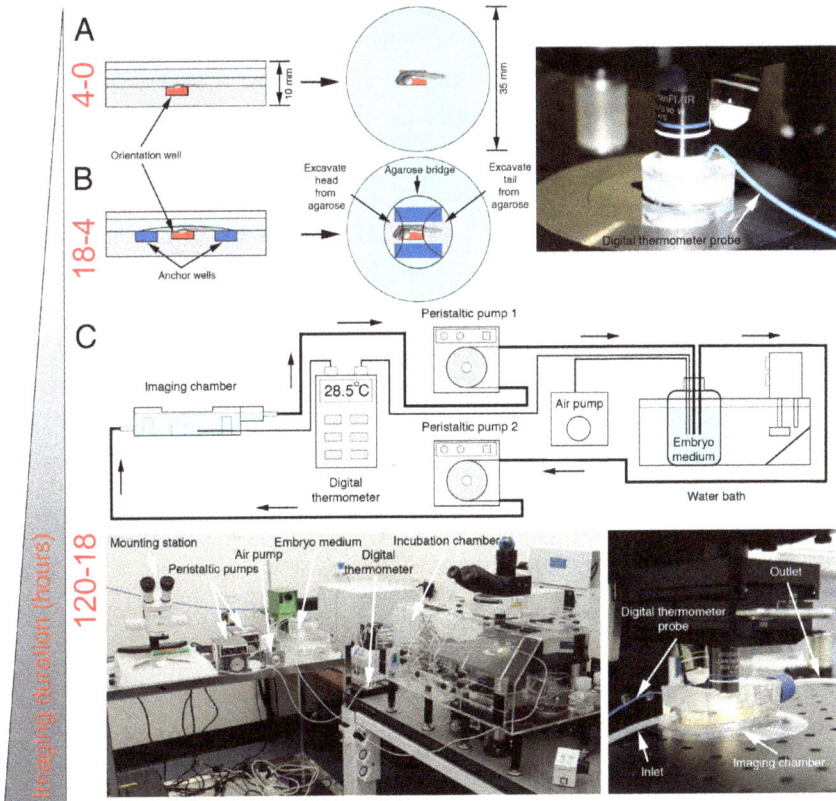

**Fig. 1** Live imaging setups. (A, B and C) Embedding and imaging setups for live confocal time-lapse imaging over zero to four, four to 18 and 18 to 120 hour imaging periods, respectively. (**A**) For imaging up to four hour duration, anesthetized embryos are embedded in a thin (~1 mm) layer of 1% lmp agarose (supplemented with PTU and tricaine) and oriented with the aid of a well (excavated from an agarose bed) to accommodate the yolk so as to ensure that the embryo remains as close to horizontal as possible within a 35 mm plastic tissue culture dish (in the case of imaging a ventral tail wound). (**B**) For time-lapse imaging spanning four to 18 hour duration, instead of overlaying the entire embryo with 1% lmp agarose, an agarose bridge is created by first placing a small amount of molten agarose over the anesthetized embryo and orienting the embryo with the aid of a well. Anchor wells either side of the embryo provide anchor points for this agarose bridge while the solidified agarose is removed from the head and tail to accommodate embryo growth and elongation. (**C**) For imaging experiments longer than 18 hour duration, embryos are mounted within a specialized imaging chamber through which temperature-controlled aerated embryo medium (supplemented with PTU and tricaine) is recirculated, as previously described.[18] All imaging is performed within a temperature-controlled incubation chamber attached to an Olympus FV1000 scanning confocal microscope.

and tail are excavated from the agarose leaving an agarose bridge pinning the embryo/larva to the underlying agarose bed. Wells either side of the embryo/larva provide points to anchor the bridge in place. This accommodates the growth of the specimen throughout the imaging period. For imaging extending in duration beyond 18 hours, we employ a temperature-controlled aerated buffer recirculating system (as previously described[18]) (Fig. 1C). In summary, a specialized imaging chamber[18] houses the embryo/larva that is mounted in a similar fashion as above (i.e. with removal of agarose from head and tail). Peristaltic pumps drive aerated E3 medium (supplemented with PTU and tricaine) from a water bath, the temperature of which is adjusted to maintain the embedded specimen at 28.5°C (as measured using a digital thermometer probing the imaging chamber), to the imaging chamber and back again in a simple circuit. All of the above strategies employ the use of an incubation chamber which is also adjusted to maintain the embedded embryo/larva at a constant 28.5°C.

## Investigating the Physiological Behaviors of Myeloid Leukocytes During Normal Development

In addition to providing a defense mechanism during early development, phagocytic cells of the innate immune system also function during organogenesis and tissue remodeling to clear the embryo of dead cells. Precisely controlled cell death is crucial for the proper formation of necessary body parts and organs during development.[19,20] Zebrafish require apoptosis for normal development[21] and represent an ideal platform in which to live image the contribution of specific phagocytic cell compartments to clearing redundant cells during tissue remodeling, for example during pharyngeal skeleton formation. During live imaging of *Tg(lyz:EGFP)* larvae, we have observed numerous marked myeloid leukocytes migrating within and around the developing branchial arches, suggesting a role for these cells in helping model these skeletal elements[14] (Fig. 2). Another system that relies upon apoptosis for correct formation is the circulatory system. Endothelial cell survival is regulated through both

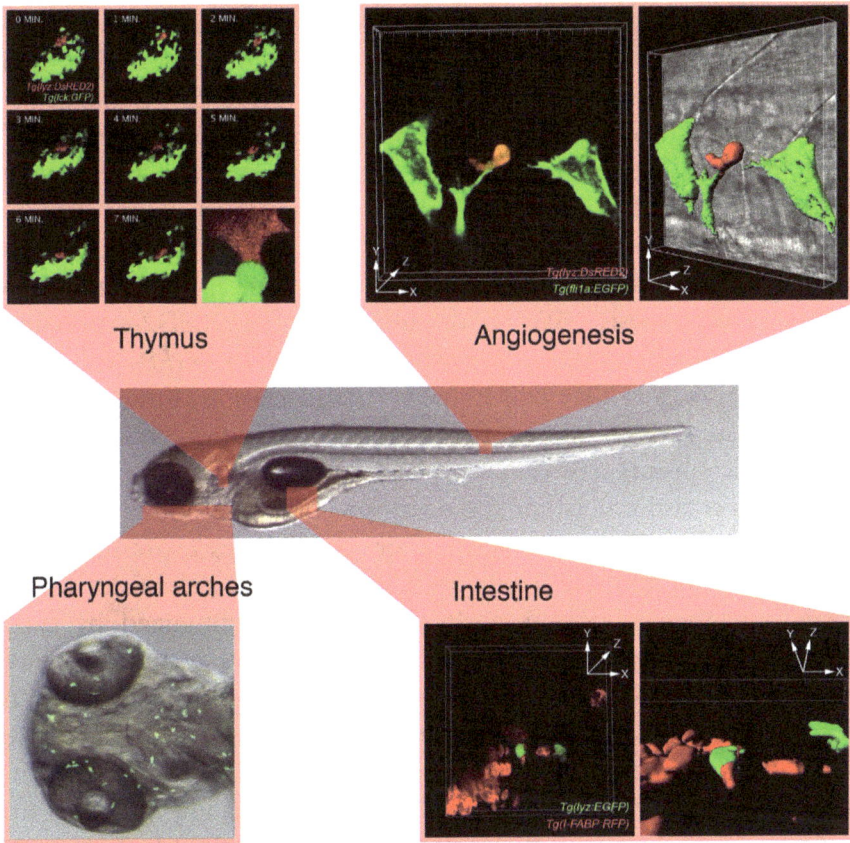

**Fig. 2** Live imaging non-inflammatory homeostatic activities of myeloid leukocytes during development. ("Thymus") Frame shots assembled from merged Z-stack projections every 60 seconds, demonstrating migration of a *lyz:DsRED2*-expressing leukocyte between marked thymocytes within the thymus of an eight dpf *Tg(lyz:DsRED2)/Tg(lck:GFP)* larva. ("Angiogenesis") Volume and isosurface renderings of confocal Z-stack through the forming intersegmental vessels demonstrating close association between a fluorescently labeled endothelial cell and a myeloid leukocyte within a 34 hours postfertilization (hpf) *Tg(lyz:DsRED2)/Tg(fli1a:EGFP)* compound transgenic embryo. ("Pharyngeal arches") Ventral view of pharyngeal arches within a three dpf *Tg(lyz:EGFP)* larva. ("Intestine") Volume and isosurface renderings of confocal Z-stack through intestine of seven dpf *Tg(lyz:EGFP)/Tg(I-FABP:RFP)* larva demonstrating a marked myeloid leukocyte migrating between fluorescent intestinal epithelial cells.

pro- andanti-apoptotic signals from their immediate environment.[22,23] Endothelial cell apoptosis is essential for both physiological and pathological vascular remodeling and is believed to play a role in the initiation and progression of inflammatory and immune diseases.[24,25] Imaging immune and endothelial cell compartments in real-time within compound transgenic zebrafish will provide an elegant strategy to assess the role of innate immune cells in clearing endothelial cells that receive both physiological and pathological instructions to undergo apoptosis. We have imaged *lyz:DsRED2*-expressing myelomonocytes making contact with *fli1a:EGFP*-expressing endothelial cells as they contribute to the formation of the intersegmental vessels, raising the possibility of communication between these angiogenic sprouts and cells of the innate immune system (Fig. 2). Apoptosis is also crucial for the development of the immune system, in particular within the thymus, where thymocytes expressing non-reactive or autoreactive TCRs are eliminated.[26,27] Zebrafish, like all jawed vertebrates, possess a thymus which is colonized by early thymic progenitors that expand and differentiate into mature T cells.[28–31] These maturing thymocytes can be live imaged within *Tg(lck:GFP)* larvae, using regulatory elements that normally drive expression of an intracellular tyrosine kinase required for TCR-mediated signaling.[32,33] We have exploited this imaging potential to monitor the behavior of fluorescently marked myeloid leukocytes in the context of fluorescent thymocytes within the thymus of *Tg(lyz:DsRED2)/Tg(lck:GFP)* compound transgenic larvae (Fig. 2). Following colonization of the thymus, we observe *lyz:DsRED2*-expressing myeloid leukocytes migrating within the thymus, appearing to make contacts with surrounding *lck:GFP*-expressing thymocytes. The physiological relevance of this interaction is unclear but is suggestive of a surveillance role for these phagocytic cells, possibly in contributing to the clearance of maturing thymocytes as part of a rudimentary selection process.

The potential to image immune cell compartments within a whole animal setting, such as that offered by the zebrafish, also allows questions to be addressed concerning the establishment and functional role of cellular immunity within different organ systems. An important interface between the host and both commensal and pathogenic microbes is the intestinal epithelial wall. From the time that the intestine exists as an open tube, at approximately four days post-fertilization (dpf), it is colonized by bacteria.[34] This necessitates a strategy for their recognition and tolerance, and in pathological settings for establishing an immune response. Very little is known regarding the establishment of mucosal immunity within the zebrafish intestine. An important strategy used within the mammalian intestine, essential for tolerance of commensal microbiota and in regulating the mucosal immune system, is that mediated through Toll-like receptor (TLR) signaling.[35] TLRs represent a highly conserved class of pattern recognition receptor that recognize specific invariant pathogen components. We have demonstrated that the zebrafish intestine, in addition to innate immune cells, expresses adaptor molecules necessary for TLR signaling, in particular MyD88, an adaptor that in mammals is required for the majority of TLR-mediated signaling.[36] A recent study has suggested that this innate defense mechanism is operational within the larval intestine of the zebrafish.[37] In this study, neutrophil influx within LPS-stimulated larval intestines was demonstrated to be at least partially dependent upon MyD88 function, as were the homeostatic numbers of intestinal neutrophils recruited in response to normal commensal microflora.[37] We have live imaged *lyz:EGFP*-expressing myeloid leukocytes migrating between intestinal epithelial cells within seven dpf *Tg(lyz:EGFP)/Tg(I-FABP:RFP)* larvae, suggesting an early immune surveillance function for these cells within the developing intestine (Fig. 2).[14] We have also addressed the potential for these innate immune cells to contribute to resolving a bacterial challenge triggered by live fluorescently-labeled enteropathogenic *Salmonella*.[14]

## Live Imaging the Response of Myeloid Leukocytes to Wounding

The inflammatory response initiated by tissue damage involves recruitment of a number of leukocytic lineages, including neutrophils and macrophages. This inflammatory response is essential to protect the host from infection and also likely has a role in guiding tissue repair processes, such as re-epithelialization.[38] The timing of myeloid leukocyte infiltration differs for neutrophils and macrophages. Typically, neutrophils invade the wounded tissues first by diapedesis from adjacent activated blood vessels primarily to destroy pathogens. Macrophages, largely derived from blood-borne monocytes, arrive slightly later, phagocytosing dead neutrophils, extracellular matrix and tissue debris as well as guiding angiogenic and further immune responses through the secretion of angiogenic mediators and inflammatory cytokines.[38] Exactly how these leukocyte lineages interact with each other and how they interpret chemoattractant and chemorepulsive factors that ultimately guide their migration throughout a wound-healing response remains unclear. Live imaging the wound response within transgenic reporter lines has already confirmed the usefulness of the zebrafish in providing novel insights into the leukocytic response.[12–16,39–41] Coupling this live, real-time, whole animal imaging potential with the genetic tractability the zebrafish system affords places this model system in a unique position to provide a level of insight into the inflammatory response not possible using more traditional mammalian model systems.

Our group uses time-lapse confocal microscopy to characterize and dissect the contribution of fluorescently marked leukocytes to wounds generated within the ventral fin. Soon after wounding, an immune response initiates within the developing larval fin, evidenced by the rapid infiltration of leukocytes to the damaged tissue.[12,14] Typically, we generate wounds in the ventral fin immediately below a 2-somite-wide region below the fifth and sixth somites distal from the cloaca (Fig. 3A). Gentle pressure is applied to the ventral-most surface of the fin using sterilized fine forceps. This creates reproducibly sized wounds that do not extend beyond ½ the length of the fin, to ensure a modest leukocytic infiltrate that enables the migratory paths of individual leukocytes to be differentiated (Fig. 3B). The characterization of novel transgenic reporter lines which mark increasingly refined leukocyte subsets will help expand our understanding of how these various innate immune cells function during the wound response. We have previously described multiple migratory responses demonstrated by

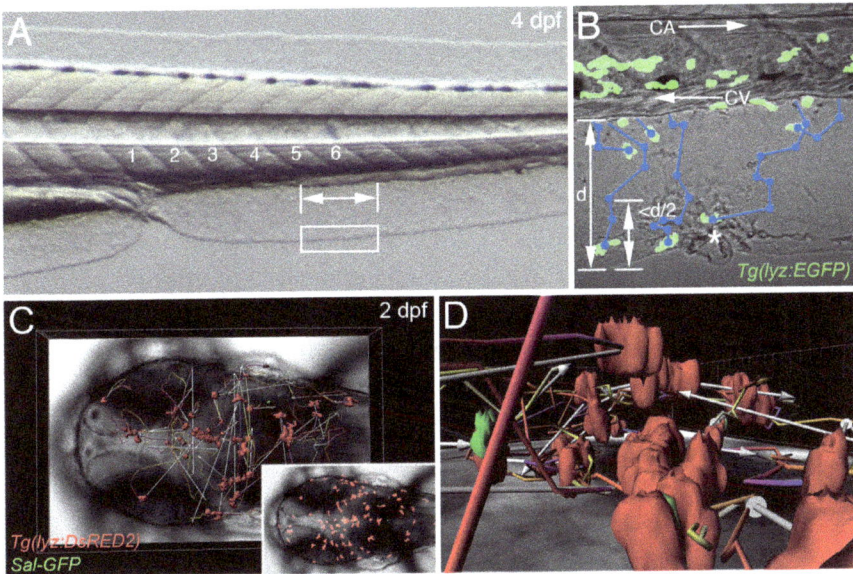

**Fig. 3** Live imaging the migratory response of leukocytes to wounding and bacterial infection within whole zebrafish larvae. (**A**) Lateral view of four dpf larval tail highlighting region of ventral fin that is wounded (ventral-most surface of fin immediately below fifth to sixth somite distal to cloaca). (**B**) A typically-sized wound, that does not extend beyond ½ the length of the ventral fin, will result in a leukocyte response of sufficient size to enable individual migration paths to be tracked. (**C** and **D**) 3D isosurface rendering from time-lapse confocal imaging demonstrating tracked leukocyte paths and displacement following injection of GFP-labeled *Salmonella* into the hindbrain of *Tg(lyz:DsRED2)* larvae (inset shows 2D Z-stack projection). (**D**) Magnified view of (**C**) reveals extensive migration of leukocytes within the Z-axis. Asterisk marks wound. CA, caudal artery; CV, caudal vein.

wound-responding leukocytes using the wounding strategy outlined above, including the ability of individual leukocytes to visit wounded tissues multiple times.[14] However, the time-lapse movies used to illustrate these behaviors, although informative, do not fully reflect the complex 3D environment in which they are occurring. Using 4D imaging software, such as Imaris (Bitplane AG, Switzerland), dynamic events occurring with respect to the Z-axis, that are not evident when constructing movies from Z-stack projections, can be revealed. Even within the relatively narrow tissue environment of the ventral fin, leukocyte movements in the Z-axis can go unappreciated when assessing their migration using Z-stack projections. Such movements in the Z-axis are more significant when imaging within thicker and more complex tissues. Following microinjection of GFP-expressing *Salmonella typhimurium* into the hindbrain of *Tg(lyz:DsRED2)* larvae, 4D reconstruction from time-lapse confocal imaging enables particle tracking algorithms to reveal extensive migratory paths of individual fluorescent leukocytes in all dimensions (Figs. 3C and 3D). Such 4D imaging will become increasingly valuable to fully appreciate the complexity of immune cell interactions that occur in all three dimensions throughout the different phases of an immune response. Its utilization will also be important for maintaining the zebrafish platform at the forefront of live real-time imaging approaches.

The use of photoactivatable caged fluorescent dyes allows the targeting of individual immune cells of known identity either before, during or following their contribution to an immune response to assess their fate (Fig. 4). We inject CMNCBZ-caged carboxy-Q-rhodamine (Invitrogen, Carlsbad, CA) dye into one-cell stage *Tg(lyz:EGFP)* embryos and at the desired developmental stage use them for ventral wounding experiments (Fig. 4A). Individual fluorescent leukocytes at the wound can then be targeted for uncaging and tracked to determine their fate. Once a leukocyte has been targeted for uncaging, it is spot-exposed to 405 nm laser pulses, typically one second exposures at 25%–30% laser intensity (Fig. 4B). The number of pulses varies depending upon the abundance of caged dye within the targeted cell. The SIM (SIMultaneous) scanning function of the Olympus FV1000 confocal system permits real-time monitoring of the uncaging progress. Uncaged leukocytes can then be detected by co-localized rhodamine and EGFP fluorescence and subsequently tracked (Fig. 4C). Time-lapse imaging these leukocytes following infiltration of the wound has revealed a number of fates. This includes the ability of some leukocytes, although initially on a migration path back to the wound to cease this trajectory and migrate in the opposite direction, despite neighboring leukocytes that initially migrate along the same path continuing their migration back to the wound

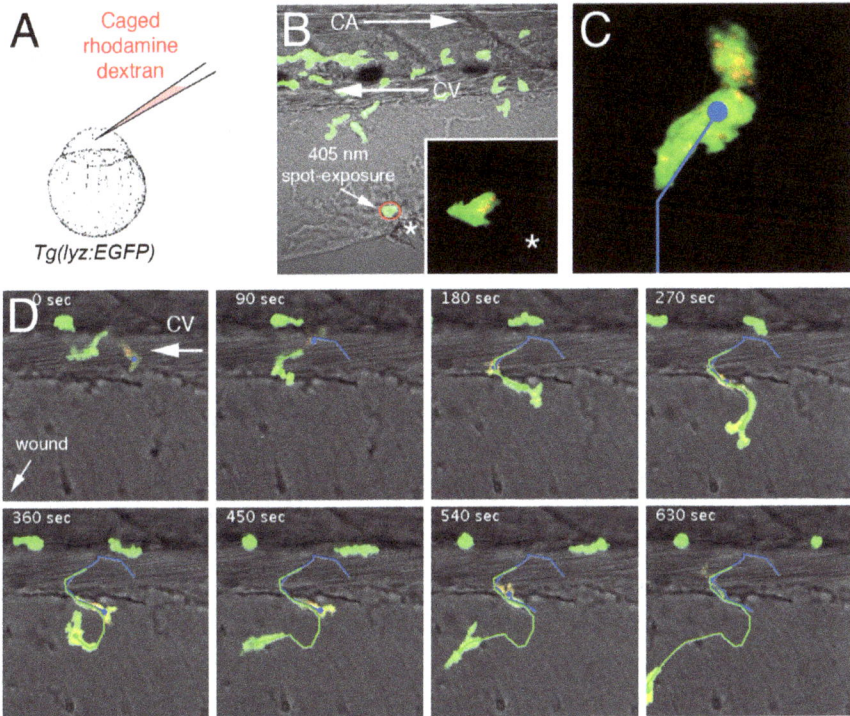

**Fig. 4** Tracking individual leukocyte responses following infiltration of wound. **(A)** Individual *Tg(lyz:EGFP)* embryos are injected at the one-cell stage with caged carboxy-Q-rhodamine then allowed to develop (protected from light) until wounding. **(B)** At desired developmental stage injected *Tg(lyz:EGFP)* larvae are wounded in the ventral fin and monitored for infiltrating leukocytes. Fluorescently marked wound-responding leukocytes are then targeted for spot exposure to a 405 nm laser to uncage red fluorescence (inset). **(C)** Following excitation, uncaged *lyz:EGFP*-expressing leukocytes can be tracked in real-time to assess their fate following initial migration to the wound. **(D)** Merged Z-stack projections from time-lapse confocal microscopy tracking wound-uncaged *lyz:EGFP*-expressing leukocytes after they have returned to the caudal vein (selected frame shots every 90 seconds). Although both leukocytes initially migrate along similar paths, one returns to the wound (green track) while the other migrates back towards the dorsal surface of the CV (blue track). Asterisks mark wound. CA, caudal artery; CV, caudal vein.

(Fig. 4D). Live imaging of these events, and ultimately dissecting their genetic determinants, has potential to enhance our understanding of the resolution of inflammation and to determine the cell-intrinsic decisions made by a leukocyte that control how it interprets the myriad of chemoattractant and chemorepulsive signals it receives from its environment. A full understanding of why leukocytes decide to either infiltrate, disperse or die during inflammation and what controls these responses has important therapeutic potential.

## Live Imaging the Leukocytic
## Response to Bacterial Infection

A number of studies have confirmed the ability of embryonic zebrafish myeloid leukocytes to contribute to inflammatory responses initiated following bacterial challenges.[7,10,11,15,42–45] To assess the potential of fluorescently marked embryonic leukocytes to mount a response to a bacterial challenge, we use fluorescently tagged live bacteria and several different fluorescent *E. coli* "BioParticles" (Invitrogen, Carlsbad, CA) that permit detection during live confocal imaging experiments. To determine the phagocytic capacity of *lyz:EGFP*-expressing leukocytes, we have adapted the Invitrogen cell culture reagent pHrodo that highlights the phagocytic potential of cells by emitting intense pH-dependent red fluorescence within the acidic environment of phagosomes. The dye is bound to either heat- or chemically-killed *E. coli* "BioParticles" to promote uptake by innate immune cells. We deliver this reagent into the circulation and then monitor its ingestion by fluorescently-labeled myeloid leukocytes. Within 30 minutes post-injection, red fluorescence begins to mark intracellular compartments within *lyz:EGFP*-expressing cells (Fig. 5A). Although these pHrodo-labeled bacteria can highlight the phagocytic potential of cells, because they only begin to emit fluorescence when within the cell they do not enable live imaging of the phagocytic event itself. The Alexa Fluor-labeled range of fluorescent *E. coli* "BioParticles" enables live real-time imaging of the phagocytosis of these fluorescent dead bacteria and facilitates the use of various combinations of

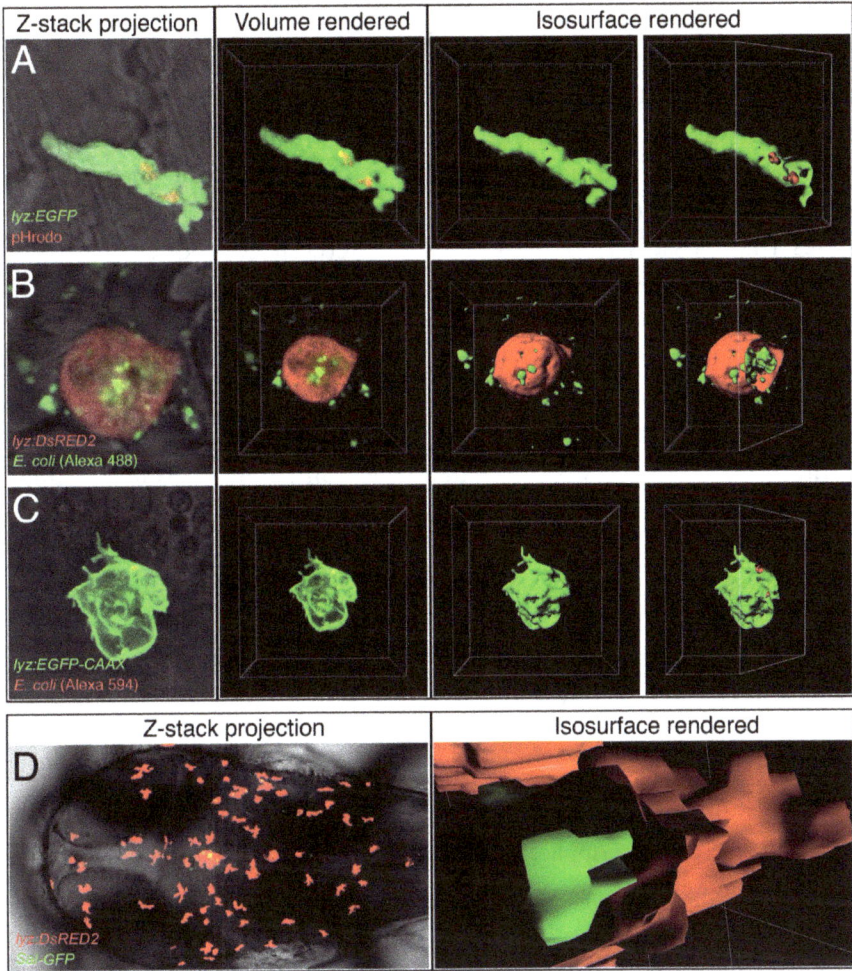

**Fig. 5** Live imaging the phagocytic potential of embryonic leukocytes. (**A**) Marked leukocyte within two dpf *Tg*(*lyz:EGFP*) embryo following ingestion of pHrodo *E. coli* "Bioparticles" to mark their phagosomal compartments with red fluorescence following ingestion. (**B** and **C**) Marked leukocytes within *Tg*(*lyz:DsRED2*) and *Tg*(*lyz:EGFP-CAAX*) embryos, respectively, demonstrating ingestion of Alexa Fluor 488-labeled (B) and Alexa Fluor 594-labeled (C) heat- or chemically-killed *E. coli*. Each image is presented as a merged Z-stack projection, a volume rendering and isosurface renderings (with a cutting plane to reveal ingested fluorescent bacteria). EGFP-CAAX, membrane-localized (prenylated) EGFP. (**D**) Merged Z-stack projection through head of two dpf *Tg*(*lyz:DsRED2*) embryo following injection of GFP-expressing *Salmonella typhimurium* and isosurface reconstruction demonstrating phagocytosis of bacterial challenge.

fluorescent bacteria and leukocytes (Figs. 5B and 5C).[36] This can then be taken a step further with the use of live fluorescently labeled bacteria such as GFP-expressing *Salmonella* (Fig. 5D).[14] The ability to live image such infection experiments within a whole animal setting as genetically tractable as the zebrafish has the potential to contribute to our understanding of complex immune-related diseases that possess both genetic and bacterial components such as inflammatory bowel disease.[46]

# Acknowledgments

The authors would like to thank the Biological Imaging Research Unit (The University of Auckland) for imaging assistance, Alhad Mahagaonkar for management of the zebrafish facility, Annie Chien and Lisa Pullin for expert technical assistance. We would also like to acknowledge Jen-Leih Wu and Nick Trede for gifting us the *Tg(I-FABP:RFP)* and *Tg(lck:GFP)* reporter lines, respectively. This work was supported by a grant from the Foundation for Research Science and Technology, New Zealand.

# References

1. Scheinecker C. Application of *in vivo* microscopy: evaluating the immune response in living animals. *Arthritis Res Ther* 7:246–252, 2005.
2. Germain RN, Miller MJ, Dustin ML and Nussenzweig MC. Dynamic imaging of the immune system: progress, pitfalls and promise. *Nat Rev Immunol* 6:497–507, 2006.
3. Willett CE, Cortes A, Zuasti A and Zapata AG. Early hematopoiesis and developing lymphoid organs in the zebrafish. *Dev Dyn* 214:323–336, 1999.
4. Lam SH, Chua HL, Gong Z, Lam TJ and Sin YM. Development and maturation of the immune system in zebrafish, *Danio rerio*: a gene expression profiling, *in situ* hybridization and immunological study. *Dev Comp Immunol* 28:9–28, 2004.
5. Zapata A, Diez B, Cejalvo T, Gutierrez-de Frias C and Cortes A. Ontogeny of the immune system of fish. *Fish Shellfish Immunol* 20:126–136, 2006.

6. Davidson AJ and Zon LI. The 'definitive' (and 'primitive') guide to zebrafish hematopoiesis. *Oncogene* 23:7233–7246, 2004.

7. Herbomel P, Thisse B and Thisse C. Ontogeny and behavior of early macrophages in the zebrafish embryo. *Development* 126:3735–3745, 1999.

8. Le Guyader D, Redd MJ, Colucci-Guyon E, Murayama E, Kissa K, Briolat V, Mordelet E, Zapata A, Shinomiya H and Herbomel P. Origins and unconventional behavior of neutrophils in developing zebrafish. *Blood* 111:4–5, 132–141, 2008.

9. Bertrand JY, Kim AD, Violette EP, Stachura DL, Cisson JL and Traver D. Definitive hematopoiesis initiates through a committed erythromyeloid progenitor in the zebrafish embryo. *Development* 134, 4147–4156, 2002.

10. van der Sar AM, Musters RJ, van Eeden FJ, Appelmelk BJ, Vandenbroucke-Grauls CM and Bitter W. Zebrafish embryos as a model host for the real time analysis of *Salmonella typhimurium* infections. *Cell Microbiol* 5:601–611, 2003.

11. Davis JM, Clay H, Lewis JL, Ghori N, Herbomel P and Ramakrishnan L. Real-time visualization of mycobacterium-macrophage interactions leading to initiation of granuloma formation in zebrafish embryos. *Immunity* 17:693–702, 2002.

12. Mathias JR, Perrin BJ, Liu TX, Kanki J, Look AT and Huttenlocher A. Resolution of inflammation by retrograde chemotaxis of neutrophils in transgenic zebrafish. *J Leukoc Biol* 80:1281–1288, 2006.

13. Renshaw SA, Loynes CA, Trushell DM, Elworthy S, Ingham PW and Whyte MK. A transgenic zebrafish model of neutrophilic inflammation. *Blood* 108:3976–3978, 2006.

14. Hall C, Flores MV, Storm T, Crosier K and Crosier P. The zebrafish lysozyme C promoter drives myeloid-specific expression in transgenic fish. *BMC Dev Biol* 7:42, 2007.

15. Meijer AH, van der Sar AM, Cunha C, Lamers GE, Laplante MA, Kikuta H, Bitter W, Becker TS and Spaink HP. Identification and real-time imaging of a myc-expressing neutrophil population involved in inflammation and mycobacterial granuloma formation in zebrafish. *Dev Comp Immunol* 32:36–49, 2008.

16. Zhang Y, Bai XT, Zhu KY, Jin Y, Deng M, Le HY, Fu YF, Chen Y, Zhu J, Look AT, Kanki J, Chen Z, Chen SJ and Liu TX. *In vivo* interstitial migration of primitive macrophages mediated by JNK-matrix metalloproteinase 13 signaling in response to acute injury. *J Immunol* 181:2155–2164, 2008.

17. Kamei M, Isogai S and Weinstein BM. Imaging blood vessels in the zebrafish. *Methods Cell Biol* 76:51–74, 2004.
18. Kamei M and Weinstein BM. Long-term time-lapse fluorescence imaging of developing zebrafish. *Zebrafish* 2:113–123, 2005.
19. Penaloza C, Lin L, Lockshin RA and Zakeri Z. Cell death in development: shaping the embryo. *Histochem Cell Biol* 126:149–158, 2006.
20. Kinchen JM and Ravichandran KS. Journey to the grave: signaling events regulating removal of apoptotic cells. *J Cell Sci* 120:2143–2149, 2007.
21. Yamashita M. Apoptosis in zebrafish development. *Comp Biochem Physiol B Biochem Mol Biol* 136:731–742, 2003.
22. Luttun A and Verhamme P. Keeping your vascular integrity: what can we learn from fish? *Bioessays* 30:418–422, 2008.
23. Santoro MM, Samuel T, Mitchell T, Reed JC and Stainier DY. Birc2 (cIap1) regulates endothelial cell integrity and blood vessel homeostasis. *Nat Genet* 39:1397–1402, 2007.
24. Clarke M, Bennett M and Littlewood T. Cell death in the cardiovascular system. *Heart* 93:659–664, 2007.
25. Winn RK and Harlan JM. The role of endothelial cell apoptosis in inflammatory and immune diseases. *J Thromb Haemost* 3:1815–1824, 2005.
26. Zhang N, Hartig H, Dzhagalov I, Draper D and He YW. The role of apoptosis in the development and function of T lymphocytes. *Cell Res* 15:749–769, 2005.
27. Feig C and Peter ME. How apoptosis got the immune system in shape. *Eur J Immunol* 37Suppl 1:S61–70, 2007.
28. Kissa K, Murayama E, Zapata A, Cortes A, Perret E, Machu C and Herbomel P. Live imaging of emerging hematopoietic stem cells and early thymus colonization. *Blood* 111:147–1156, 2008.
29. Murayama E, Kissa K, Zapata A, Mordelet E, Briolat V, Lin HF, Handin RI and Herbomel P. Tracing hematopoietic precursor migration to successive hematopoietic organs during zebrafish development. *Immunity* 25:963–975, 2006.
30. Jin H, Xu J and Wen Z. Migratory path of definitive hematopoietic stem/progenitor cells during zebrafish development. *Blood* 109:5208–5214, 2007.
31. Langenau DM and Zon LI. The zebrafish: a new model of T-cell and thymic development. *Nat Rev Immunol* 5:307–317, 2005.

32. Trede NS, Zapata A and Zon LI. Fishing for lymphoid genes. *Trends Immunol* 22:302–307, 2001.
33. Langenau DM, Ferrando AA, Traver D, Kutok JL, Hezel JP, Kanki JP, Zon LI, Look AT and Trede NS. *In vivo* tracking of T cell development, ablation, and engraftment in transgenic zebrafish. *Proc Natl Acad Sci USA* 101:7369–7374, 2004.
34. Bates JM, Mittge E, Kuhlman J, Baden KN, Cheesman SE and Guillemin K. Distinct signals from the microbiota promote different aspects of zebrafish gut differentiation. *Dev Biol* 297:374–386, 2006.
35. Abreu MT, Fukata M and Arditi M. TLR signaling in the gut in health and disease. *J Immunol* 174:4453–4460, 2005.
36. Hall C, Flores MV, Chien A, Davidson A, Crosier K and Crosier P. Transgenic zebrafish reporter lines reveal conserved Toll-like receptor signaling potential in embryonic myeloid leukocytes and adult immune cell lineages. *J Leukoc Biol* 85:751–765, 2009.
37. Bates JM, Akerlund J, Mittge E and Guillemin K. Intestinal alkaline phosphatase detoxifies lipopolysaccharide and prevents inflammation in zebrafish in response to the gut microbiota. *Cell Host Microbe* 2:371–382, 2007.
38. Martin P and Leibovich SJ. Inflammatory cells during wound repair: the good, the bad and the ugly. *Trends Cell Biol* 15:599–607, 2005.
39. Mathias JR, Dodd ME, Walters KB, Rhodes J, Kanki JP, Look AT and Huttenlocher A. Live imaging of chronic inflammation caused by mutation of zebrafish Hai1. *J Cell Sci* 120:3372–3383, 2007.
40. Redd MJ, Kelly G, Dunn G, Way M and Martin P. Imaging macrophage chemotaxis *in vivo*: studies of microtubule function in zebrafish wound inflammation. *Cell Motil Cytoskeleton* 63:415–422, 2006.
41. Cvejic A, Hall C, Bak-Maier M, Flores MV, Crosier P, Redd MJ and Martin P. Analysis of WASp function during the wound inflammatory response — live-imaging studies in zebrafish larvae. *J Cell Sci* 121:3196–3206, 2008.
42. Clay H, Davis JM, Beery D, Huttenlocher A, Lyons SE and Ramakrishnan L. Dichotomous role of the macrophage in early *Mycobacterium marinum* infection of the zebrafish. *Cell Host Microbe* 2:29–39, 2007.
43. Prajsnar TK, Cunliffe VT, Foster SJ and Renshaw SA. A novel vertebrate model of *Staphylococcus aureus* infection reveals phagocyte-dependent resistance of zebrafish to non-host specialized pathogens. *Cell Microbiol* 10:2312–2325, 2008.

44. Davis JM and Ramakrishnan L. The role of the granuloma in expansion and dissemination of early tuberculous infection. *Cell* 136:37–49, 2009.
45. Brannon MK, Davis JM, Mathias JR, Hall CJ, Emerson JC, Crosier PS, Huttenlocher A, Ramakrishnan L and Moskowitz SM. *Pseudomonas aeruginosa* Type III secretion system interacts with phagocytes to modulate systemic infection of zebrafish embryos. *Cell Microbiol* 11:755–768, 2009.
46. Xavier RJ and Podolsky DK. Unravelling the pathogenesis of inflammatory bowel disease. *Nature* 448:427–434, 2007.

# Index

1-phenyl 2-thiourea (PTU)
    132–134
2A peptide    51

autocorrelation function (ACF)
    72, 73, 76

blood flow velocity 86
branchiomotor neuron    1, 3, 4
    facial branchiomotor neuron
        (FBMN)    3–13

central nervous system (CNS)    1,
    17, 18, 20–23, 25, 26, 28–30
centrosome    51, 54, 56–62
cerebellum    38–42, 48, 58, 60
cholesterol    106, 109–114, 120
confocal    21
    microscopy    5, 6, 11, 138, 141
Cre/LoxP system    13
cross-correlation function (CCF)
    77, 78, 93, 94

differential interference contrast
    (DIC) microscopy    131
diffusion coefficient    70–73,
    75–77, 81–83, 88, 90, 92, 95
digital scanned laser sheet
    microscopy (DSLM)    12

dislipidemias    106
dissociation constant    70, 71, 83,
    92–94

E. coli "Bioparticles"    142, 143
enteric neuron    114

facial nerve (nVII)    3
flila:EGFP    135, 136
fluorescence correlation
    spectroscopy    69–71
fluorescence resonance energy
    transfer (FRET)    78, 95

Gal4    48, 49, 52–54
Gal4/UAS system    13
gall bladder    112–114, 118
glia    18–22, 25–27, 29
    perineurial    26, 29
    peripheral    17, 25, 27, 29
glial    17–20, 29, 30
glial-glial    19
G-protein coupled receptor
    (GPCR)    71, 90
granule cell    48
green fluorescent protein (GFP)
    1, 4, 5, 7, 9, 13
growth cone    37
guidance    36, 42, 60

healing  129, 138
hindbrain  1, 3–5, 8, 12, 13

*I-FABP-RFP*  135, 137, 144
ImageJ  9
Imaging
  4D  139, 140
  bio-imaging  37, 47, 52, 62, 63
  live  129–135, 137–139, 142, 143
  software  8
Imaris  8, 139
immunity  136, 137
*in vivo* cell biology  36, 37, 43, 46, 55, 58, 63
infection  129, 130, 138, 139, 142, 144
innate  129–131, 134, 136–138, 142
intestinal enterocyte  110
intestine  106–108, 110–113, 115, 118, 119
Islet1-EGFP  89, 90

larvae  107–109, 112–117, 119, 120
leukocyte  130–132, 134–144
lipase  105, 106, 109, 111, 116–119
lipid  105–115, 119–121
  drops  109
  metabolism  105, 107–109, 111–114, 119, 120
  transport  106, 109
lipolysis pathway  109

*lyz:DsRED2*  135, 136, 139, 140, 143

macrophage  131, 137, 138
medial and lateral ganglionic eminence  38
membrane-bound GFP (EGFPCAAX)  7
midbrain-hindbrain boundary (MHB)  40, 41
molecular interaction  70, 77, 78, 92
myeloid leukocyte  130–132, 134–137, 142

neural cell adhesion molecule (NCAM)  41
neurogenesis  19, 20
neuronal migration  1, 2, 35–39, 41–43, 46, 49, 50, 54–56, 59–63
neurulation  19, 20
neutrophil  131, 137, 138
Nipkow disk  44, 45
nucleokinesis  56–60, 62

observation volume  69, 72–79, 81–86, 91
olfactory system  38
oligodendrocyte  20–22, 24, 25
oligodendrocyte progenitor cell (OPC)  22–24, 29, 30
one-photon excitation  71
organelle dynamics  43, 59, 61

pancreas  108, 118, 119

pathway 3, 5
PED6 112–118
perineurial 25, 26, 28–30
perineurium 25, 26, 28
peripheral nervous system (PNS) 17–21, 25–30
photo-convertible fluorescent protein 45
planar cell polarity (PCP) 3–5, 7
prickle 5
progenitor cell 21
protease 105, 115, 116, 118, 119

resonant frequency or "Eigenfrequenz" 45
Rohon-Beard mechanosensory neuron 19
rostral migratory stream (RMS) 38

*Salmonella typhimurium* 140, 143
Schwann cell 19, 20, 25, 26, 28–30
selective place illumination microscopy (SPIM) 11, 12
sensory neuron 19
single molecule event 70–73, 96
single-wavelength fluorescence cross-correlation spectroscopy (SW-FCCS) 69, 70, 74, 77, 80, 81, 92–94
slit-scanning confocal microscopy 45
spinning disk confocal microscopy 44

subventricular zone (SVZ) 2
surgical trauma 130

*Tg*(*β*-actin:HRAS-EGFP) 13
*Tg*(*Hras:GFP*) 9
*Tg*(isl:GFP) 4, 5, 9, 13
*Tg*(lyz:EGFP) 134, 135, 140, 141, 143
*Tg*(*zic4:GFP*) 48
time-lapse 18, 20–24, 26, 27, 30
transgenic 17, 18, 22, 26, 30, 130, 135, 136, 138
triacylglyceride 109, 111
tricaine 132–134
two-photon excitation 77, 78
two-photon microscopy 1, 5, 6, 8, 9, 11–13

upper rhombic lip (URL) 40, 41, 58

very low density lipoprotein (VLDL) 109
Volocity 8

Wnt-signaling 4
wound 129, 130, 132, 133, 137–141

yolk 108, 109, 114

zebrafish 17–21, 25, 26, 29, 30, 35–43, 46–53, 55, 57–63, 69–71, 81–88, 90, 92, 93, 96, 130–132, 134, 136–140, 142, 144